THE ROLE OF SCIENCE IN CIVILIZATION

THE ROLE

OF

SCIENCE

IN CIVILIZATION

by ROBERT BRUCE LINDSAY

GREENWOOD PRESS, PUBLISHERS
WESTPORT, CONNECTICUT

Library of Congress Cataloging in Publication Data

Lindsay, Robert Bruce, 1900–
    The role of science in civilization.

    Reprint of the 1st ed. published by Harper & Row,
New York.
    Bibliography: p.
    1. Science and civilization. I. Title.
[CB151.L5   1973]   901.9   73-3234
ISBN 0-8371-6837-6

Originally published in 1963 by Harper & Row, Publishers,
New York

Reprinted with the permission of Harper & Row, Publishers, Inc.

Reprinted in 1973 by Greenwood Press, Inc.,
51 Riverside Avenue, Westport, Conn. 06880

Library of Congress catalog card number 73-3234
ISBN 0-8371-6837-6

Printed in the United States of America

10 9 8 7 6 5 4 3 2

CONTENTS

Chapter 4

SCIENCE AND PHILOSOPHY

Chapter 5

SCIENCE AND HISTORY

Chapter 6

SCIENCE AND COMMUNICATION

PREFACE

No one questions the importance of science in our contemporary civilization. Unfortunately, the nature of science and its relations to other ways of describing and understanding human experience are not well understood by many people, including the representatives of the humanities in our culture. The present book was written in the attempt to make a small contribution toward the clarification of this problem. It has developed from a university course which the author has given at Brown University for the past five years, its purpose being to encourage students to strive for a better integration of their specialized knowledge in many fields.

A book of this kind inevitably owes a great debt to many more sources than can be specifically acknowledged. The bibliography at the end of the volume will indicate to some extent the sort of material the author has found of value in stimulating his own thoughts. Certain parts of Chapters 6 and 9 have been taken from the author's article in the September, 1959, issue of the *American Scientist* ("Entropy Consumption and Values in Physical Science") by kind permission. The author wishes to acknowledge his indebtedness to Henry Margenau for permission to use and discuss in Chapter 9 certain unpublished material of the latter.

Thanks are due to D. Van Nostrand Company, Inc., for permission to use Fig. 6.1, and to the Springer Verlag for permission in connection with Fig. 6.3 and Table 6.2.

It is perhaps fair to call attention to the index, which in addition to subject references contains an annotated list of names (with dates) of practically all persons referred to in the text.

The author is deeply indebted to Mrs. Susan Greenwood of the Department of Physics, Brown University, for her painstaking typing of a large part of the manuscript and for valuable assistance in the preparation of the index.

R. B. L.

CHAPTER 1

THE SETTING

OF THE PROBLEM

What do we mean by the role of science in civilization and what excuse can we offer for studying it? When the subject is mentioned, most people simply call attention to the host of mechanical gadgets which are so obvious a part of life in the so-called Western world and are becoming ever more conspicuous even in so-called backward countries; they imply that nothing more need be said. We take our electric power for granted, with all the appliances which depend on it for their operation, and feel utterly disorganized when for some reason it fails. More and more we become dependent on the internal combustion engine for transportation, but are not surprised to see jet propulsion cut long distance travel times to figures that would have seemed almost inconceivable a half century ago. Is this what science means to life: a way of introducing devices to make living more complicated, more exciting, more varied, more comfortable—but also more dangerous?

Leaving aside for the moment the validity of this estimation and accepting it on its face value, we must admit that it provides plenty of justification for a thoroughgoing appraisal of the impact of science on the life of man in society. For it presents a host of problems, of which the most fundamental is: can society safely permit the uncontrolled application of science to all aspects of life? The frightening potentialities of nuclear energy come to mind at once in this connection. One might well raise the question: by what right do scientists and engineers force on society an energy source which by very virtue of its magnitude can, if misused, produce evil on the same large scale as good? Such questions are ethical and not primarily scientific in character, but they are inextricably connected with the relation between the results of science and the concomitant human behavior. Hence, they must be faced in any discussion of science and society.

But let us go back to the initial question and the common reaction to it. The obvious is often misleading just because it *is* obvious. However, it is precisely the not-so-obvious influence of science on civilization which has in the long run changed our lives far more than the readily perceptible technological applications. What is the nature of this nontechnological role? It is a bit difficult to name it, since it has not been formally baptized. We might call it the philosophical or ideological influence, or the role of science by virtue of its ideas. This may strike some people as bizarre, since it is not uncommon to meet the belief that science does not involve *ideas*: it deals with *facts*. This, however, betrays a very limited understanding of the real nature of science. Science does contain and do business with vast quantities of factual knowledge. But it could never organize this without the guidance of concepts which are freely created by the imaginations of men. These concepts change with time and have much to do with the way in which people interpret their experiences.

Consider in this connection the notion of "common sense." Aristotle considered it was common sense to say that a heavy object falls to the ground because it seeks its proper place, as close to the center of the earth as possible, while a light object rises because its proper place is above the earth. At the same time he felt it was common sense to describe motion in general by dividing it all into two classes, natural and violent. Proceeding in this way he produced what was to him and his contemporaries a satisfactory classificatory account of motion. This, however, did not seem to be common sense to Galileo, who felt that some new ideas were necessary if we are to understand motion. To him the notion of inertia, i.e., the tendency of an object in motion to stay in motion unless interfered with, seemed more in keeping with common sense. This change in the notion of common sense could hardly have come about without the intervention of scientific activity. It is part of the ideological influence of science on our daily lives.

Or take another closely connected example of change in the concept of common sense. We have mentioned that the Aristotelian Greeks and Europeans for many centuries after them felt it was eminently sensible to think of the falling object as seeking its proper or natural place. After the development of the Newtonian theory this no longer made sense. Rather, the released object falls to the earth because it is attracted by the earth with the force of gravitation. The impartial observer might object that it is difficult to see why one view is any more sensible than the other. This would be true were it not for the fact that the Newtonian hypothesis of gravitation is part of a highly developed theory from which one can deduce consequences amenable to experimental test. The tests have in general been remarkably successful. The impact of such a procedure on human thinking about nature has been considerable. It marks a complete change in view as to what constitutes the understanding of experience. It provides confidence to man that he can find regularities or order in experience, and not merely order postulated *ad hoc*, but order that he can do something with, i.e., on the basis of which he can predict future experience. This change in scientific thinking had incalculable effects on all other thinking in the seventeenth, eighteenth and nineteenth centuries. It stimulated rationalism in philosophy, political theory and sociology. And it helped to release the ordinary man from the bonds of superstition.

Though the example we have given is taken from physical science, the influence of scientific concepts has of course not been confined to this field. In biology everyone is aware of the tremendous role which the theory of evolution has played in changing earlier views of man's nature and origin. And this in spite of the fact that the theory and its predictions are by no means as clear-cut as the theories of physics.

It must be emphasized that we need make no claim here that the ideological influence of science on society has always been beneficial, though this is the natural attitude of scientists. All we are trying to do is to recognize the existence of the influence. The appraisal of its worth will appear later in our discussion. This is a rather important point, since much of the hostility to science exhibited by humanistic scholars is based on a feeling that science has somehow undermined the finer values of life and by its influence is in a fair way to destroy civilization not so much through technology as through its materialistic ideas and the use of a methodology utterly at variance with that used by the humanities in handling human experience. One of the purposes of this book is to make clear that this feeling of the humanist is based on a profound misunderstanding of the nature of science and that a better appreciation of the role of science in our society will mitigate the unhappy suspicions and antagonisms. This is vitally necessary if intellectuals are to join ranks in attacking with all the resources of which human beings are capable the problems of society.

It should be obvious that in embarking on any serious study of the role of science in civilization the first step is to set forth clearly what science is. This we proceed to do in Chapter 2 with illustrations drawn mainly from physics but with references to other branches of science as well, to emphasize the fundamental identity of purpose and method throughout. This will be followed by an examination of the alleged differences between science and the humanities, in which it will appear that a more careful and candid approach will turn up overriding similarities which are generally overlooked. No evaluation of the role of science in society can afford to neglect the relations between these two important ways through which human beings cope with their experience.

Philosophy as a humanity is in a rather special position vis-à-vis science, and therefore demands particular attention. We go into this in Chapter 4. This leads on directly to the perennial question of the relation of science to the social studies, particularly history. Is history a science? If not, does it have scientific aspects? What of the history of science? What does it have to do with the role of science in civilization? We shall try to find out.

Probably the most important single element in the influence which any method of coping with experience has on the development of civilization is communication. This therefore demands considerable attention, especially in view of the enormous strides that information theory is now taking and the great contemporary interest in cybernetics and the general problem of control. Language and its structure are being looked at in more scientific fashion with consequences which may prove fateful for the human race. All educated people need to understand more clearly the role of mathematics in communication.

Though science and technology are not the same, their relation is vital for mankind, and hence we devote a chapter to tracing it. The vanishing frontier between basic and applied science is one of the most significant aspects of our contemporary civilization. Hence, it is essential to examine the development of technology from early times, even if in a rather summary fashion in order to understand the role which it has played in the advancement of science as well as the reciprocal influence which has been exercised on it by science, especially in modern times. Basic scientific research performed in the industrial laboratory is a twentieth century phenomenon of tremendous significance for society.

The importance of technology for the conduct of war brings the state into the scientific picture and suggests that the relations between science and government be exposed. This we do in Chapter 8 and call attention to the significance of the very large financial support which the state now provides for both pure and applied science. The problems raised thereby with respect to freedom in science assume a foremost place. For as soon as science serves the state on a large scale, the problem of power enters the picture, and the privilege of the individual to go his own way in uncommitted fashion can become seriously affected. This in turn may have a vital influence on the progress of science.

Finally, no account of the role of science in modern society could claim any vestige of completeness if it did not consider the problem of the relation of science to human behavior, especially on the plane of ethics. We therefore devote the last chapter in the book to asking whether there can be a science of ethics, or failing this whether successful scientific concepts can provide any useful analogies in the field of ethics. Our conclusion will be on the conservative side, and yet we shall find in the theory of thermodynamics elements of interest from the ethical standpoint and shall present a thermodynamic imperative.

Here in brief is what the book proposes to do in the way of illuminating the role of science in civilization. Let us get at the task, first examining the nature of this science which we are talking about.

CHAPTER 2

WHAT IS SCIENCE?

Definitions

Brief definitions of a form of human intellectual activity as elaborate and comprehensive as science are notoriously difficult and perhaps of questionable value. Yet the reader is entitled to some glimpse of the thing he is being asked to think about before he plunges into the details. Let us note at the beginning a few of the many attempts to epitomize science.

It has been called *organized knowledge,* especially in high school texts. But we are aware that much knowledge is acquired and displayed at that well-known sporting institution, the race track, and this knowledge is admittedly highly organized. It would be wrong to call it science, however. The definition, if meaningful at all, includes too much.

A well-known modern philosopher has defined science as the "search for the perfect means of attaining any end." This again seems rather too broad: the criminal in search of the perfect crime is hardly a scientist necessarily!

Attributed to Albert Einstein is the statement: "The whole of science is after all nothing but a refinement of everyday thinking." This may encourage all those who believe they do "everyday thinking" into believing they are scientists, and this includes practically everybody! But alas, the step involved in the word "refinement" is a considerable one, as anyone who has taken the trouble to look at the technical works of the distinguished author of the statement will readily attest.

The best way to find out what science is, is to examine with care what scientists do. Though their activities are extremely diverse in detail, depending on the particular disciplines to which they give attention, leading some to believe that after all science is not subject to an all-embracing characterization, close inspection shows that this attitude is too despairing. One can, after all, come fairly close to epitomizing science as follows.

Science is a method for the description, creation and understanding of human experience.

Let us look into this a bit. In the first place we must make clear what this human experience we are talking about really is. By it we shall mean the sum total of all the sense impressions of human beings together with their mental reflections on these sensations. An enormous amount of philosophical thought has developed on the meaning of experience. We shall not stop to appraise this here, though it will receive some attention further on in the book. At the moment we assume merely that human beings have various ways of coping with their experiences, that knowledge is possible and can be shared. Science is one way of grappling with experience, so as to make some sense out of it. It is of course not the only way: the artist handles experience in a different fashion, though perhaps the fundamental difference between him and the scientist has been exaggerated, as we shall see in the sequel. Science should, at any rate, be humble and not claim that it alone knows the way.

What then do we mean when we call science in the first instance a method for the description of experience? Confronted by the ceaseless flux of sensation we naturally tend to seek for some order, some pattern in what at first appears as mere chaos. Certain sense impressions are recognized as repeating themselves, e.g., the alternation of day and night, or the recurring feelings of hunger or sleepiness. This periodicity is undoubtedly one of the first things seized on by the human observer as indicating an order in experience, something on which he can rely and to which he can adjust his own activities with assurance. In addition to such recurrences of phenomena, interesting relations continually turn up connecting apparently diverse bits of experience. Thus, for example, originally both electricity and magnetism seemed to be quite separate phenomena. Electricity was associated with the peculiar ability of certain materials like amber, rubber and glass when rubbed by appropriate substances to attract or repel light objects like paper. Magnetism likewise was found to exhibit phenomena of attraction and repulsion but only with certain metals and metallic ores. Ultimately (but not until the nineteenth century A.D.), it was discovered that electricity in motion produces magnetic effects, and from magnetism in motion one can produce electrical effects. Relations of this sort are part of the order of our experience which discerning individuals are constantly on the lookout for. What more natural than that they should proceed to talk about such things with their fellows! This is the first step in description, a simple account in ordinary language of such regularities of experience as passive observation turns up. That such things happen is clear enough from the historical records of the race, though this by no means traces out the enormous difficulties that must have stood in the way of the process for primitive man, not to mention the terrifically complicated psychological problems involved in an attempt to understand how it came about that our ancestors ever attempted such a descriptive procedure.

It is customary to say that curiosity motivated and still motivates the human tendency to describe experience. From time immemorial people have been curious as to "how things go," and must have realized at an early stage that the first step in grasping the significance of any phenomenon is to describe it clearly and accurately so that there may be a common comprehension of it among so-called normal persons. Perhaps it would be fairer and more true to the facts to say that some people are curious in an out-of-the-ordinary sense and have such a passionate desire to see more deeply into experience that they devote more of their time and attention to an examination of what really goes on than the bulk of the population cares to. Not everyone has cared to be a scientist, nor is the urge uncommonly conspicuous today. But there always have been people who have sought to find relations like that noted above connoting regularity among apparently different portions of experience, as for example, lightning and thunder, the tides and the phases of the moon, the seed and the flower, electricity and magnetism, the appearance of an organ in the living organism and its physiological functioning, etc. Such relations, when found to have sufficient generality, are known as scientific *laws*, and we shall examine their nature and meaning in some detail further on.

One should not overlook another urge to the description of experience, the desire for *power*. This takes at least two forms: firstly, the realization that careful observation and recording of certain aspects of experience can provide a measure of control. Thus, observation leading to more or less precise knowledge of the effect of wind on the sails of a ship ultimately provides a means of transportation by sea. Description of the behavior of the stars leads to a system of navigation at sea. Description of experience confers power over experience, as man has learned from primitive times onward. In the second—and less attractive—form of this search for power, man has sought to turn his precise knowledge of experience into power over his fellow human beings. This is evident in all ages of human advance. The military expert takes advantage of his superior knowledge to create more potent weapons which enable him to bring his foes into subjection. The person who has discovered the medicinal powers of herbs and other preparations holds the power of life and death over those who consult him. The astronomer who can accurately predict a solar eclipse can terrify a whole population. We hardly worry about such matters nowadays in the fashion of our ancestors to whom such knowledge was akin to magic. But what shall we say of those who applied knowledge of nuclear fission and fusion to the development of the atomic bomb? Is it an exaggeration to say that they have created an instrument of incalculable power over the destinies of persons yet unborn? We shall return later to a detailed consideration of the power problem in science.

Experiment: The Creation of Experience

We have not yet done justice to science as description of experience. To do so necessitates an examination of the second important element in the method, namely, the creation of experience. This takes place through the agency of experiment. When people decided to supplement passive observation by the active acquisition of experience a great forward step was taken in the scientific endeavor. This involves in brief a conscious arrangement of physical objects in the hope that something new will emerge. The result has often been termed *controlled* sense perception, but it is really more than this: it is genuine creation of experience that would not have otherwise come to light.

A little analysis of the fundamental character of experiment will serve to bring out its essential importance. Suppose an investigator wishes to perform an experiment. What must he do? In the first place, he must delimit the region of experience he desires to examine. He may wish to study, for example, the properties of matter under the influence of heat. He must therefore abstract from the sum total of experience the small domain involving the heating of objects and give his attention to that, trying as far as possible to exclude all other effects. This abstraction and concentration on a limited field of investigation is the first and obviously a very important characteristic of the experimental method: one can hardly perform an experiment on the universe!

The second important characteristic of an experiment is careful planning on the part of the experimenter. Clearly, he must have some ideas in his mind as to what he wishes to do and why and how he proposes to set about doing it. In the case of the effect of heat on an object, for example, he may decide to find out what happens to the color of a metal on heating. This involves some expectation that the color of the heated metal will be different from that at room temperature and a feeling of curiosity as to what the change actually is. To carry out his intention, a source of heat must be planned and a suitable group of metallic specimens procured. A series of steps must be organized, for the scientist early recognizes that nothing much is to be gained by merely aimless procedures. All this means that a great deal of preliminary thought goes into any scientific experiment and the more sophisticated the state of the science the more elaborate is the planning, involving as it often does a complicated design of apparatus in order to realize as far as possible in actual fact the abstraction in the mind of the experimenter and to insure effective "controls."

It is often said to be an essential attribute of a scientist that he approach the performance of an experiment with an open mind. In so far as this means that he will record results honestly and without bias or prejudice, this is of course a correct statement. But it cannot mean that he has no preconceived ideas—these he must have or he would never perform the experiment at all.

Once having planned the experiment, the scientist must take the final step: he must go through the necessary operations. These are important and often demand considerable manual dexterity, but since they can often be carried out by a competent mechanic under the guidance of the planner, they do not constitute so fundamentally significant a part of the whole concept of experiment as the first two characteristics. Some very great experimenters have been rather inept technicians and manipulators.

The essentially arbitrary character of experimentation must not be overlooked: the scientist *chooses* the kind of experience he desires to create. This aspect of science is often misunderstood by those who take the attitude that science is the study of nature as something existing outside man and awaiting discovery. Actually, the scientist consults his own interest in picking fields for examination and experimentation. He may indeed be guided by accidental observations—and often is—but he has to exercise choice in his procedures and in a strict sense his discoveries are governed accordingly.

Measurement

The concept of experiment as set forth in the preceding discussion contains no necessary quantitative aspect. One can think of many scientific experiments involving only qualitative procedures and results. But in certain sciences, notably those usually called *physical* and to a growing extent in all science, experimentation has not been considered really significant unless it involves somehow an answer not only to the question "how" but also to "how much." As soon as this quantitative urge enters the picture, experiment becomes measurement.

The answer to the question "how much" obviously entails attaching numbers to the results of an experiment. Instead of merely noting that the sound of a distant bell reaches our ears after we see the hammer strike, we want to know *how long* it takes the sound to travel the relevant distance. Instead of merely noting that of two objects, one seems heavier than the other as we hold them in our two hands, respectively, we wish to be able to say that the one is so many times heavier than the other. There is thus an urge to assign numerical significance to observations and our experiences generally, particularly those which refer to the behavior of so-called physical objects.

Measurement can be as simple as counting the number of petals on a flower or as complicated as determining the constant of gravitation, i.e., the constant which enters into Newton's law of gravitation, giving the force with which any two particles in the universe attract each other. In the physical sciences, and indeed in all science in so far as it follows the lead of physics, the fundamental measuring device is the *scale*. In its simplest form this is a suitable physical surface on which a set of marks is inscribed in some definite, prearranged fashion. A meter stick is of course an illustration, as is also the face of a clock or a thermometer. Attached to the marks are numbers arranged in increasing order of magnitude. The spacing of the marks is arbitrary, though the most common type of scale is a linear one, in which the number assigned to any particular mark is directly proportional to the distance of the mark from the chosen origin (assigned the number zero). There are of course other important scales, of which the logarithmic is an example, its purpose being to compress into a relatively small space a large range of values of a given quantity.

To use a scale for a measurement we need to establish a coincidence between a mark on the scale and some physical object such as a pointer. Thus, in the case of the clock dial there are two pointers, namely, the two "hands." In an electrical meter, on the other hand, there is a needle which can move along the scale. The number attached to the mark on the scale with which the needle or pointer coincides yields the result of the measurement. Of course, it is clear that many conventions involving nomenclature, units, etc., must be arbitrarily established before meaning can be given to the numbers resulting in this way. With these we are not concerned at the moment. The reader is urged only to think of all the measuring instruments that have come within his ken and note that they all involve a scale and a pointer in some form or other. Small wonder that A. S. Eddington referred to all the results of physical measurements as "pointer readings."

The use of scales in measurement evidently implies a careful consideration of properties of space and geometry which we can hardly go into here. Rather more important in the understanding of measurement is the meaning of symbolism, since every measurement is ultimately described in symbolic terms. Evidently a merely operational description, saying that when such and such equipment was arranged with a scale and pointer, certain readings were obtained when certain conditions were varied, would be of little value. One has to *talk* about measurements and indeed experiments in general in a language reflective of the meaning one desires to attach to them in the context of the whole domain of experience to which they refer. This can only be done in an efficient manner by the introduction of an appropriate symbolism.

What is a symbol? In its most general sense, a symbol is an object representative of a definite situation which may result from either observation or contemplation. The marks on paper or the sounds we utter as speech, constituting human language, are of course examples of symbolism—a powerful aid in human communication. Of course, an object like a flag can also be a symbol, but in science the symbols employed are in general those of language, though usually, and in particular in the physical sciences, they are of a more abstract character than those of ordinary speech. The association of symbols with numbers makes quantitative communication possible, and the association of symbols with classes of objects or with propositions is the basis of mathematics and symbolic logic. We shall discuss this later at greater length. Here we are more concerned with the use of symbolism in the description of experiment and measurement.

Symbolism in Experiment and Measurement

In the early stages of any branch of science when the descriptive aspect is predominant and experiment is on a purely qualitative level, symbolism may consist merely of the introduction of a linguistic terminology, a set of names attached to certain appearances or operations. The names given to the parts of the body are an example. Much of classical biology employs symbolism of this character. Its usefulness is largely associated with the collection and classification of data.

When measurement comes into the picture the situation changes. Let us examine an illustration from the physical measurement of gases. When one decides to observe what happens when a gas is squeezed, he has to talk about what he does in such a way that another person can go and do likewise. Moreover if he desires to introduce a quantitative aspect into his work, he must invent something to measure. A squeeze implies change in volume and something to which the change in volume may be thought to be due. So the experimenter devises a set of operations which he calls measurement of volume.

This is not in general an easy thing to do, particularly if the vessel in question has an irregular shape. For regular vessels one falls back on simple solid geometry, but other containers demand more elaborate techniques, such as liquid displacement, etc. Ultimately, all such methods must fall back on geometry for justification and thus can lead to some difficult mathematics.

The something to which the change in volume is thought to be due is denominated *pressure,* and a set of operations for assigning values to this quantity is also required. We do not go into details, but the reader will recall various kinds of so-called pressure gauges, such as mercury tube manometers, ionization gauges, etc. The point is that a set of operations is introduced which ultimately leads to the assignment of a number to denote the magnitude of the squeeze to which the gas in the container has been exposed. Instead of describing these operations in each case in detail we find it convenient to use the abstract symbol p for the result of the measurement and say we are measuring pressure. Thus, there is a tremendous gain in economy of language in the use of the abstract symbolism. But more than that, we can treat the symbols p and V, representing the operations leading to the measurement of pressure and volume, respectively, as *algebraic* symbols which can have mathematical relations with each other and which can therefore represent routines of experience. In the special example being discussed, such a relation might be

$$p\,V = \text{constant}$$

for constant temperature. That is, the interpreted content of all the operations of squeezing, together with the assignment of numbers to the symbols p and V representing these operations, reduces to the algebraic statement that the pressure and volume vary inversely at constant temperature.

Before we could reach the above conclusion, we should, of course, have to plot the results of each individual measurement by indicating in a diagram (Fig. 1.1) with a cross the measured value of the pressure for each volume. This gives a set of crosses in the so-called p V plane. The next step is to draw the smooth curve which passes as near to each cross as possible. The mathematical equation of this

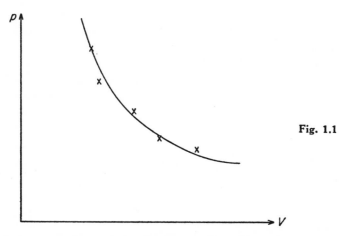

Fig. 1.1

curve is found to be the aforementioned equation. Many readers will recognize it as the expression of what is called Boyle's law in physics.

Before we could have much confidence in this equation as an abstractly symbolic representation of the set of operations on the gas, we should naturally have to repeat the operations for varied conditions. If in all such cases the mathematical form of the result stays the same, we are permitted to conclude that we have a description of a regularity of experience, which we may call a *scientific law*. Though the abstract symbolism of mathematics is certainly not necessary for its statement, it has proved of enormous value in securing economy of expression as well as in drawing precise conclusions.

The Concept of Scientific Law

A scientific law is a symbolic description of a routine of experience. The use of the word "law" in this connection must be carefully distinguished from its use in common speech, where it can have many meanings from "common" law through "statute" law to "moral" law, all of which possess connotations that have little or no relevance to the use of the term in its scientific sense. For example, scientific law has nothing in common with a law established to force persons to avoid certain behavior. Likewise, it has nothing to do with the so-called "necessity" of divine law. Probably the very term should have been avoided in scientific nomenclature, but it has now become so well established that it will not readily be relinquished.

We have already mentioned Boyle's law as a good illustration of a law in physics. The student of physics will recognize other examples in Hooke's law for the behavior of elastic materials, Ohm's law for the behavior of electric currents in conductors, Snell's law for the refraction of light, etc. The chemist has his law of definite proportions, Henry's law for the solubility of a gas in a liquid as a function of the partial pressure of the gas, etc. The biologist will recall the Mendelian law of inheritance, the law of growth of organisms, etc. In psychology we have the Weber-Fechner law connecting stimulus and response in the human being and the law of association of ideas. Even in the social field, laws have been established. Such a one is that connected with the name of Gresham and having to do with the circulation of money (bad money drives out good money). Other laws will be mentioned in later chapters.

It is not pretended that all these laws have the same degree of validity or precision. In general, those in the physical sciences are subject to expression in terms of the abstract symbolism of mathematics as indicated previously, and numbers may be assigned to the symbols by measurement which can be made highly precise. As we proceed to the life and social sciences, the expression of laws becomes more vague. This need not necessarily always be the case; clearly, the effort in every science is for definiteness and explicitness and the removal of ambiguities.

There are a few other features of the scientific law that we ought to note. For one thing, its range of application is limited in each case. Thus we must not conclude that Boyle's law applies universally to all gases under all conditions. It does not apply, for example, to carbon dioxide below $31\,°C$. Hooke's law does not describe the behavior of solids which are stressed so strongly that they are in danger of breaking. Hence, in the statement of a law it is strictly speaking necessary to state the precise conditions under which it is expected to serve as a good description. Few physical laws are found to hold over wide ranges of conditions; this appears to contradict the expectation of some earlier scientists and philosophers, of whom Newton was an outstanding example, that the more carefully experience is studied, the simpler will scientific laws become. Of course, much depends here on what is meant by simplicity!

Simplicity is a deceptively attractive criterion but is apt to elude our grasp when we try to come to grips with it. In particular, it seems to be primarily a function of familiarity and this is largely a matter of individual education; a person who approaches calculus for the first time often finds the concepts difficult to grasp. After some practice and experience, however, he is willing to say that the calculus terminology is simple. The attempt has indeed been made to interpret simplicity in terms of the minimum number of independent concepts needed for the expression of a law, but this itself is somewhat vague, since it is often difficult to disentangle concepts and establish clearly the meaning of independence. One might be tempted to say that a monotonic mathematical function is simpler than an oscillatory one, but physicists have got so used to sinusoidal functions and find them so helpful that this distinction is illusory.

It was at one time customary to look upon scientific laws as representing the inexorable course of nature, as having a fundamental element of necessity about them. This view is no longer held. Laws are descriptive only and the realms in which they describe accurately are limited, as we have already noted. More precise measurements may reveal their inadequacy. Thus the Weber-Fechner law for the relation between stimulus and response in psychology was originally expressed in terms of a logarithmic function, but more recent research has shown that a power law expresses the experimental observations better.

The Concept of Scientific Theory

We have discussed science as a method for the description and creation of experience. We now come to what is probably the most important characteristic of science, namely the *understanding* of experience. This represents the urge to go behind the description to find some meaning in the laws, to endeavor to answer the question *why* instead of being satisfied with the question *how*. The scientist strives to understand by creating theories. What is a theory?

A scientific theory is a picture formed by the human mind, of such a nature that from it may be deduced by logical processes of thought an array of laws which can be identified with those experimentally verified. The picture thus created can be closely related to an everyday experience or it can contain purely imaginative elements. Let us illustrate with an example from physics.

Suppose the domain we wish to understand more deeply is that of the behavior of gases subjected to changes of volume, temperature and pressure. Now gases are fluids and as such seem to be continuous in nature so far as ordinary sense perception is concerned. But it is hard to deduce many detailed properties from a mere assumption of continuity. One can indeed apply the principles of mechanics to such a medium and thus establish a theory of hydrodynamics, but this would provide no reason why the fluid can be made to occupy a smaller volume by squeezing it. By the same token it would not be able to predict that such a squeeze is propagated as an elastic wave with a definitely calculable velocity. It is not then surprising that the physicist has sought to provide fluids with a structure which is not obvious to the senses but which is created by the mind. This structure is the content of the so-called kinetic theory of gases, which is basically a picture of the fluid as collection of a vast number of discrete particles called molecules, moving in all directions with a wide range of velocities and colliding frequently with each other and the walls of the containing vessel. This is of course a variety and extension of the atomic theory, which is as old as the Greeks.

It will be noted that the picture involves a series of hypotheses. The basic idea is that fluids are to function as if the picture were real. In other words, it is hoped that logical deductions from the theory will agree with experience, and indeed that the theory will predict new results which have never been observed before. The testing of such new predictions is an important part of the creation of experience.

We cannot stop here to work out the consequences of the kinetic theory of gases. This involves applying the principles of mechanics to the motion of the molecules and also the introduction of statistical ideas, since the molecules are too numerous for an exact solution of the motion of each one. We can at any rate see in a general way how the theoretical picture might lead to observed properties. In the first place, it does not require too much imagination to see that the continual bombardment of the walls of the container by the molecules exerts a force on them which taken per unit area can provide a theoretical representation of the quantity measured as pressure.

Moreover, the greater the number of impacts per unit time the greater the pressure if we assume a kind of average velocity for all the molecules. It is also evident that when the volume of the container is decreased (as by the motion of a tight-fitting piston) while the number of molecules and their average speed remain the same, the number of impacts per second must increase, corresponding to an increase in observed pressure. It is therefore not hard to see how Boyle's law can be deduced from a kinetic theory picture. Even the idea of temperature can be worked into the picture without trouble. For the preceding argument depends on the assumed constancy of the average speed of the molecules. If we further assume that it is the average kinetic energy (one half the mass times the square of the speed) of the molecules which corresponds to what we measure as temperature on a thermometer, constancy of average molecular speed means constancy of temperature. Hence the significance of constant temperature in the usual statement of Boyle's law is assured. Naturally, the precise functional relation in the expression of the law depends on the detailed formulation of the hypotheses, and this means the use of mathematics.

If the kinetic theory of gases were limited in its predictive power merely to the deduction of Boyle's law, we might be mildly interested but certainly not impressed. The real merit of the point of view inherent in the kinetic theory is its ability to predict some property of a gas, some regularity in its behavior not hitherto observed. The kinetic theory has fulfilled this obligation. One of the fascinating episodes in the history of the physics of gases is the development by James Clerk Maxwell in the middle of the nineteenth century of a theory for the viscosity of a gas based on the molecular model. It will be recalled that viscosity represents the tendency on the part of any fluid to resist the relative motion of its parts. A very simple illustration is found in the tendency of a metal disc rotating about an axis normal to its center in a gas or liquid to come to rest after it has been given an initial angular velocity and even though the friction of the supports is negligible. It is this property of a fluid also which accounts for the fact that when a fluid flows through a pipe there is a fall off in the flow speed from the center to the periphery. It is possible to express this viscous drag in terms of a coefficient, called the coefficient of viscosity, defined to be the ratio of the tangential drag per unit area between two parallel flowing layers of fluid divided by the gradient of the flow velocity normal to the flow direction. From the kinetic theory Maxwell was able to derive an expression for the viscosity coefficient of an ideal gas as a function of other standard properties of the gas, such as temperature and pressure. He found the somewhat surprising result that whereas the viscosity coefficient does depend on the temperature and indeed increases directly as the square root of the absolute temperature, it is independent of the pressure. No experimental work existed with which to compare this prediction. Hence Maxwell himself and the German physicist O. E. Meyer proceeded to measure the viscosity of air over a range of pressures. The theoretical prediction was found to be verified. This was a triumph of the theory. To be sure, nature does not permit man to become too cocky over such success. It has turned out that if one pushes the pressure high enough the independence predicted by Maxwell's simple theory no longer holds. At high pressures of the order of many atmospheres the viscosity increases slowly. At very low pressures (e.g., in so-called vacuum) the viscosity falls off. Fortunately, the theory can be patched up so as to account for these discrepancies. The prediction of the dependence of viscosity on temperature was also felt to be at the time a rather remarkable result, since it is precisely opposite to the behavior of the viscosity of liquids, which decreases with increase in temperature, as everyone knows from common experience. But this prediction has also been verified.

The remarkable success of some scientific theories in predicting experience naturally provokes admiration and stimulates a desire to look further into the nature of such theories. Our investigation might be directed along two paths. In the first place, we might raise the question: how is it possible for some people to develop out of their heads such clever ideas as the basic assumptions of a theory? In the second place, we might seek to understand the logical structure of a theory to see how it functions. The first path would lead us into what may truly be called the psychology of science or scientific invention, while the second contents itself with an objective examination of how a theory is constructed. The former is of course far more difficult than the latter. We shall have something to say about it later. The next section takes up the logical nature of a theory with illustrations.

The Logical Structure of a Scientific Theory

For convenience, we shall analyze the structure of a scientific theory under the following five categories and then discuss each in turn.

1. Primitive, intuitive, undefined notions
2. More precisely defined concepts or constructs
3. Postulates connecting the constructs
4. Deductions from the postulates, i.e., derived laws
5. Experimental verification of the deductions.

Strictly speaking, category 5 need not be included in the logical schema, since nothing forces the scientist to verify his predictions. However, he usually does so and hence it is more satisfactory to round out the analysis in this way.

1. Let us now try to put a little flesh on the skeleton, realizing indeed that to do a complete job would require a whole book by itself. Category 1 is in some respects the most difficult of the lot. But it does not take much contemplation of any scientific theory to realize that there are some ideas which must be intuitively or instinctively accepted before we can do business at all. In any physical theory, for example, these are the fundamental but undefined notions of space and time. These are taken as part of the primitive experience of the human race. Philosophers have spent ages in trying to understand them. Scientists are usually content (though not always!) to accept them as something given. Perhaps further illustration will help to clarify what we are talking about. Everyone knows that geometry, now considered a branch of mathematics, started out as a physical theory basic for the understanding of the measurement of land. As systematized by Euclid, it begins with notions which must be accepted but cannot be further defined, such as *point, line, plane.* It is true that the endeavor is usually made to identify such words by statements such as: "A point is that which has no parts." But these are pious illusions, since the word "part" is as difficult to define without the use of prior spatial concepts as "point" itself. The customary procedure here is to fall back on human sense perception and the experience of people.

All scientific theories must begin with such primitive notions. Thus, every biological theory must necessarily start with the idea of "life," something to be recognized but undefined in the logical sense. It is proper to say that the scientist should endeavor to cut the number of such primitive ideas to the barest minimum con-sistent with carrying out his intentions in the building of his theory. It is obviously an important function of the philosophy of science to scrutinize carefully all such undefined terms in any scientific the-ory. This is one of the (to the scientist) uncomfortable parts of the business. It must be so to anyone engaged in the providing of defi-nitions. Dictionary makers must feel the same discomfort, since a dictionary is one vast tautology; every word in it must be defined in terms of other words in it. We do not, however, for this reason consider a dictionary useless.

2. Proceeding now to category 2 we feel on more assured ground, since here we begin to put the undefined terms to work and create new and somewhat more precise concepts. Since these are literally constructed by the scientist, it is perhaps reasonable to term the re-sults "constructs," as has been proposed by Henry Margenau.[1] In mechanics we begin to construct ideas like that of the material par-ticle as the fundamental entity that moves; like displacement, ve-locity and acceleration of the particle, and eventually concepts like mass, force, momentum and energy. In the kinetic theory of gases, which was introduced in a general way in the previous section, we have to build the construct of a molecule, something like the mate-rial particle, but with the possibility of greater complexity such as might be associated with two or more particles closely tied together in some way. We have to endow the molecule with mechanical prop-erties like mass and momentum, and so on. If the theory is a biologi-cal one, like the gene theory of heredity, it is necessary to define what we shall mean by a gene in terms of the chromosome structure of the cell. In the theory of evolution it is manifestly requisite to define a species. These well-defined constructs are the working entities of each theory, and in general it turns out that the success of the theory varies directly with the precision with which the definition is car-ried out. Much of the power of physical theories is directly connected with the precise agreement among physicists as to what is meant by the terms like mass and energy which they employ, and by the same token, a lot of the difficulties encountered in putting such theories

---

[1] See *The Nature of Physical Reality* (New York: McGraw-Hill Book Company, 1950).

on a firm foundation have resulted from carelessness in the defining of terms. It is clear that if the quantity mass means one thing to one investigator and something else to another, progress will be by no means assured. The history of science provides many examples of this sort of thing.

Carelessness and lack of precision in the definition of constructs are obviously less excusable in the physical sciences than in the biological sciences. For one can expect to pin down the meaning of something which is to be measured by an instrument whose construction and operation can be described in such detail that no interested, intelligent individual can fail to follow the directions. On the other hand, the specification of a unique construct in a living system runs up against the great variety of detail characteristic of such a system. Nevertheless, the successful building of theories in the life sciences demands the closest possible approach to precision that is attainable.

The problem of the definition of scientific constructs is, of course, by no means so simple or straightforward as some of the preceding statements might suggest. It has been well handled by Margenau[2] with particular reference to physics, though what he has to say can apply appropriately to all branches of science. We limit ourselves here to recalling his distinction between epistemic (or operational) and constitutive (or theoretical) aspects of definition. It is customary to emphasize that a definition of a construct to which numerical significance can be assigned, i.e., a quantity, is of little value unless it specifies the operations one must go through to measure it. Thus, a definition of mass, which was at one time common in textbooks, as the amount of matter in a body, is operationally or epistemically meaningless. On the other hand, all useful constructs in science must contain in their definitions some reference as to how they enter into the scientific laws. It is not enough, for example, to define time in physics by its measurement with a clock; one must also specify how the symbol t which stands for time enters into the mathematical equations of mechanics and other branches of physics. This is its constitutive meaning. There may indeed be some constructs which can be defined only in constitutive terms. In the first stages of the kinetic theory, this is true of "molecule." All we can say of it is how

[2] *Ibid.*, Chapter 12.

we intend to use it in developing the theory. Later, operational significance is given to properties of the molecule or aggregates thereof. A construct like the radius of an atom may never get direct epistemic significance, and yet it can have great theoretical value in the equations of atomic physics.

3. We come now to what is in many respects the nub of the whole problem of theory building, i.e., the choice of the basic hypotheses. These are the statements which the scientist assumes he can make about the concepts he has constructed. It might be supposed that if his theory is going to deal only with constructs which are defined operationally, he need make no hypotheses whatever, but merely carry out measurements of the various quantities which he thinks are related and establish the relations. But things are not quite so simple as that. In the first place, this procedure may merely lead to the establishment of a scientific law (e.g., the law of Boyle), which is of interest as a description of a regularity of experience, but whose explanation demands a theory, as we have seen. In the second place, if he is dealing with quantities defined constitutively for the molecules of the kinetic theory, for example, he is unable to carry out the measurements at all, since all he can make experiments on is gases as a whole, i.e., what may be called macroscopic phenomena. Hence, he is forced to make some statements which are impossible of immediate and direct verification by experiment, similar, for example, to the postulates of geometry. Thus Euclid had to introduce statements such as: it is always possible to draw a straight line from any point to any other point, it is always possible to describe a circle about any point as center and with any distance from that center as radius, things equal to the same thing are equal to each other, the whole is greater than any of its parts, through any point not on a given straight line it is possible to draw one and only one line parallel to the given line (the so-called parallel postulate), etc., etc. These are among the hypotheses of the theory of geometry as a scientific theory. It is true that some of them were considered by the Greeks as self-evident, as somehow forced on the mind in such a way that it is impossible to do business without them. These were termed *axioms* by Euclid. But the fact remains that he still set them up as preliminary assumptions; there seems to be no way to deduce them from more primitive statements. Others in the above list were

evidently considered even by Euclid himself in a different category. This is notably the case with the parallel postulate, which generations of mathematicians tried to show was not really a postulate at all but should be derivable from the rest of the postulates. The attempt failed, as is well known, and led to the development of non-Euclidean geometries.

So we conclude that postulates are necessary in the construction of any scientific theory. A few more illustrations will help to clarify their nature. Reverting to physics, we realize now that the basic postulates of the kinetic theory of gases are found in the assumptions that molecules exist, that they have certain properties and that the observed behavior of gases is the statistical result of this behavior. It is of course necessary to express this behavior in terms of the principles of mechanics. Hence, in order to work out the kinetic theory it is necessary to fall back on the theory of mechanics, itself one of the fundamental theories of all physics. It has become so much a part of technology, to be sure, that it is often not felt to be a theory at all, but rather something that just is! Careful analysis shows, however, that it has the same logical structure as any other physical theory.

Its primitive concepts are naturally those of space, time and motion, and its derived constructs those of displacement, velocity, acceleration, mass and force. What about the postulates? These are the famous laws of motion of Newton, the statements which connect mass, acceleration and force. Note that they are called "laws," but actually they are not laws in the sense of Boyle's law, etc. They are not statements describing routines of experience. They are really hypotheses suggested indeed by experience, but in no wise merely a description of laboratory experiments. They purport to have a generality which far transcends any specific experience of motion— with which indeed all possible motions anywhere must somehow agree. By inserting different types of force functions into the famous second "law" ($F = m\,a$) and integrating the resulting equation, it is possible to derive various actual laws of motion, e.g., the law of falling bodies, $s = \frac{1}{2}\,g\,t^2$, the law of pendulum motion, etc.

4. But this brings us to the fourth category in the logical schema, namely, the use of the constructs and postulates to derive conclusions which can be compared with laboratory experiments and measure-

ments, or what we have called scientific laws. In the preceding discussion of the nature of scientific law we looked upon a law as a description of an observed regularity of experience, an experimentally grounded statement containing concepts, all of which have operational or epistemic significance. We are now faced with a second interpretation of law, namely, as a statement deduced by logical processes from the hypotheses of a theory.

In the case of a physical theory, the process of deduction is almost certain to be mathematical. Thus to deduce the law of falling bodies: $s = \frac{1}{2} g t^2$, where s is the distance fallen in time t by a particle dropped from rest, and g is the so-called acceleration of gravity (approximately 980 cm/sec^2), the procedure involves substituting for the force F in the equation of motion above the quantity, m g, where m is the mass of the particle. This is supposed to represent the gravitational force of the earth on a particle of mass m. The resulting equation $a = g$ is really an ordinary differential equation of the second order (i.e., $d^2s/dt^2 = g$) whose solution, for the special case in which the origin of the displacement coordinate is the point from which the particle is dropped and the initial velocity is zero, is precisely $s = \frac{1}{2} gt^2$. In general, deductions from the principles of mechanics involve the solution of differential equations.

The derivation of the laws of geometry from the axioms and postulates is also a mathematical process usually known in geometry texts as the proof of the theorems. The latter are indeed the laws, e.g., the statement that the sum of the interior angles of any plane triangle is equal to two right angles. This of course is also an experimental law which can be established to any desired degree of accuracy by measurement of a great variety of triangles with a protractor. Probably the teaching of geometry is better approached by this experimental method. Nevertheless, the logical deduction of such a result produces a great deal of esthetic satisfaction in view of the generality, which the experimental technique can obviously never equal.

The method of derivation of geometrical laws or the proof of geometrical theorems certainly seems different from the derivation of the laws of moving particles. In the latter case, one operates with abstract symbols which are assumed to obey certain properties of combination and to which numerical values may be assigned without disturbing these properties. In the former, one reasons with diagrams containing features with properties assumed in the postulates. One can of course associate algebraic symbols with quantitative aspects of the geometrical figures and transform geometrical derivations into algebraic ones as in so-called analytical geometry.

To take another example, the deduction of the laws of heredity from a biological theory like the gene theory involves the use of statistical mathematics. It is largely a matter of counting permutations and combinations with the use of probability. To be sure, certain assumptions about the fate of the genes in cell division must be made to account for the two fundamental laws of Mendel, namely, the law of segregation and the law of random assortment of inherited characters. The derivation is therefore not made in such straightforward mathematical fashion as in the case of physical laws. Nevertheless it is a logical deduction.

5. The last category of the logical structure of scientific theory is the stage of verification by experiment. We have already commented on the fact that in a strict sense this does not belong to the "logical" structure of theory, but since deduction and prediction of laws are rather empty in science without experimental test, we find it convenient to comment on this aspect here.

We obviously encounter a serious problem. Before we can test the prediction of a theory it is essential to identify operationally each construct that enters into the deduced law. A theoretically derived statement whose concepts are defined constitutively only is not susceptible of experimental test. For example, before we can make a prediction of the kinetic theory mean anything in terms of the observed properties of a gas we must interpret something in the theory as pressure of the gas. Fortunately this is not too difficult since the average change in momentum of the molecules at the walls of the containing vessel lends itself very well to this sort of interpretation. The theory of mechanics thus provides both constitutive and epistemic versions of the definition of pressure. It is not always so easy to make the identification between the constructs of a theory and the operational quantities we must measure to test the theory experimentally. Thus, for example, historically it took a good deal of thought to produce the bridge between the theory of electrostatics and the experimental behavior of charged conductors and insulators. It was only when George Green in 1828 introduced the concept of the potential function that the identification became effective. Similar problems were encountered with electric currents where the new idea of electrical resistance had to be created and instruments devised to measure electric potential and current.

In light and sound the concepts of wave length and frequency had to be invented and operationally defined before the laws of the theories of these phenomena could be verified.

Comments on Scientific Theory. Criteria of Success

The brief outline of the nature of a scientific theory given in the previous section inevitably raises many questions. Such might be, for example: When can we say that a scientific theory is a true representation of experience; are the hypotheses of a theory free creations of the human mind or are they forced on the mind by experience and/or the way we have to think; does science progress toward a small number of very general theories which will forever encompass all experience? Closely connected is the question, what is the role of imagination in the framing of a theory? Finally, of what use is a theory anyway except for its esthetic appeal? This list could be readily amplified, but the discussion of these questions will, at any rate, tie in very appropriately with the problem of the relation of science to other forms of coping with experience.

The first question is a very important one, since there is a common view that science seeks the ultimate truth, and hence in this view a theory should be testable as to its contribution to truth. Unfortunately the question is really unanswerable until we decide on the meaning of truth. This is a difficult philosophical problem and there never has been a satisfactory disposition of it. It might be tempting to conclude that if the derived statements from a theory agree with experience, the theory is true. However, there is no guarantee whatever that further deductions from the theory will so agree and there are many illustrations of cases in which some conclusions of a theory have been verified and others have not. Shall we say then that the theory has been "disproved" or shown to be untrue? It may well be that the theory has been very useful in suggesting a point of view which if appropriately modified can lead to better agreement with experience. Hence, we are hardly justified in discarding it altogether.

The more one examines this matter of the "truth" of a theory, the more persuaded one becomes that the use of the word and the confused complex of meanings and misunderstandings behind it is inappropriate in science. It is far better to be satisfied with a more modest criterion of the "goodness" of a theory and talk about its *success*. If the conclusions of a theory can be reasonably identified with observed phenomena it will be called to that extent successful. This does not mean that it will not in the course of time be superseded by another theory in better agreement with experience. In fact, the whole course of science is the history of the evolution of theories.

An even more decisive criterion of the success of a scientific theory is its ability to predict experience not hitherto extant. We have already commented on this. The greater the domain of experience which a theory encompasses the more satisfied the scientist is that he understands what is going on. But his satisfaction is greatly enlarged if the theory itself leads directly to the creation of new experience by predicting something not hitherto observed and hence suggesting new experimentation. The prediction of internal conical refraction in the passage of light through a crystal by Sir William Rowan Hamilton, and its subsequent experimental verification by the Irish physicist Lloyd, was a tremendous stimulus to the wave theory of light and its subsequent development. Similarly, the prediction by Dirac from his quantum mechanical theory of the electron of the existence and properties of the positron or positively charged electron and the subsequent isolation of this particle by Anderson strengthened enormously the scientific position of quantum theory. The famous predictions of Einstein's theory of gravitation are well known. The attempts to verify them have formed a considerable part of modern astronomy and astrophysics. The derivation of the phase rule from Gibbs' theory of thermodynamics of heterogeneous substances had a tremendous influence on the development of physical chemistry.

The theory that disease in man and other animals is due to the action of microorganisms is a very old one, and there are vestiges of it in the writings of Renaissance physicians, e.g., Girolamo Frascatoro. It was predictions from this point of view which led to the famous experiments of Louis Pasteur on the preservation of food, and the combating of the diseases of animals. These predictions were also closely associated with Pasteur's experimental test and disproof of the theory of spontaneous generation.

Even theories whose predictions fail to meet the test of experiment can be very helpful in the progress of science. Thus, the corpuscular theory of light in the form adopted by Newton predicts that the velocity of light will be greater in a medium like water than in the air. On the other hand, the wave theory of light predicts that the velocity will be smaller in water than in air. This naturally stimulated experimental research in the endeavor to decide between the two points of view. The famous experiments of Fizeau and Foucault in France around 1850 decided in favor of the wave theory. However, as we shall see later, this did not entirely demolish the corpuscular theory. It only served to discredit the Newtonian formulation. In any case, the theoretical predictions led eventually to new knowledge and thus served to justify the theorizing.

One can multiply instances indefinitely in all fields of science. In the biological domain, Ernst Haeckel's famous biogenetic "law" (really principle or hypothesis) that "ontogeny (the development of an individual) is a brief and rapid recapitulation of phylogeny (the development of the race)" stirred up an enormous amount of activity in embryology. The behavioristic theory in psychology which insists that all we can ever know about human beings is by observing their behavior, and that nothing can be learned about so-called "mental" operations by introspection, produced a tremendous controversy in this science by its interpretation of psychological experience.

It must indeed be confessed that prediction as the main criterion for success of a theory is at the moment much more decisive and persuasive in physical science than in biological science. From the standpoint of the physicist this merely is a sign of the scientific immaturity of biology. The situation is rapidly changing.

The Role of the Imagination in Scientific Theorizing

One can hardly hope to understand the nature of a scientific theory without an appreciation of the character of the hypotheses which play such a key role in its development. Just how do these hypotheses arise? This is a question in what may be called the psychology of scientific invention, where the word invention is used deliberately to refer to the construction of theories. This branch of psychology has received relatively little attention and our knowledge of the processes of creative scientific thinking is rather slight.[3] We can be fairly sure that this ability for scientific invention comes in general only to those who have made themselves masters in their fields of study through experience and by profound and long-continued thought concerning this experience. But it remains true that not all who have studied and meditated thus intensively make great discoveries. Something additional is necessary, and this on the whole has eluded observation. One trouble is that too few great scientists have bothered to set down in writing their own understanding of their thinking about their novel ideas. The result usually is that one can see the beautiful structure of theory but not understand how it came to be built—the scaffolding has been removed and the plans have disappeared. Naturally scientists differ much in their methods in this respect. Michael Faraday, for example, kept a diary and recorded in it his current thinking about the problems he was trying to solve. It is possible through this to get a glimpse of the development of his ideas. With Willard Gibbs, on the other hand, the situation is different. We have some polished masterpieces of concise exposition in his *Equilibrium of Heterogeneous Substances* and his *Elementary Principles in Statistical Mechanics,* but very little about the way he came to his ideas. Of course, the attempt can be made to elucidate this by a careful reading of the published material and in fact a two-volume commentary has been prepared on the writings of Gibbs. This certainly clarifies much of his exposition, but it is doubtful whether it gets to the root of his method of procedure.

It is, of course, possible that no one is ultimately in a position to pinpoint the origin of his scientific ideas. The usual experience is that such things arise rather vaguely in the context of one's study of other men's work or the examination of one's own experimental results. By the time they get sharp enough to do business with, the author may well be at a loss to know how much he contributed and how much really came out of his background study. It is true that much has been made by some scientists of the flash of illumination at an unexpected moment. Poincaré reports somewhere that a fun-

[3] See, for example, J. Hadamard, *Psychology of Invention in the Mathematical Field* (Princeton University Press, 1945); R. Taton, *Reason and Chance in Scientific Discovery* (London: Hutchinson, 1957); A. Moles, *La Création Scientifique* (Geneva: Éditions René Kister, 1957).

damental notion in the theory of functions came to him in a moment of insight just as he was stepping into a carriage. He admits that he had meditated deeply on the problem previously, but had in the meantime put it out of his mind. The obvious psychological conclusion is that his subconscious mind had been working on the problem and the result ultimately was transferred from the subconscious to the conscious mind. Unfortunately, this provides no sure recipe for scientific creation.

We can at any rate describe the process of scientific creation in terms of a metaphor. We can say that images are somehow formed by the mind, which may be pictures of hypothetical objects, with properties something like those of observed objects, as, for example, the molecules of the kinetic theory. They may however take the form of relations among symbols which have operational significance, as in the case of the equations of the electromagnetic field. In either case, we say that the imagination is at work. This represents the ability to create a picture containing a set of entities with such properties that if they were really to exist the resulting phenomena would be those actually observed in the world of experience. It must not then be forgotten that the world of scientific theory is a construction of the human imaginative power.

It would be too long a story to analyze in detail the role of the imagination in scientific theorizing, but we ought at any rate to note the contrast between the inductive-deductive point of view of theory presented in our discussion and the straightforward inductive scheme made popular by Francis Bacon and commonly stressed as "the" scientific method. It is tempting to quote the famous passage in the introduction to his *Instauratio Magna*.

> The capital precept of the whole undertaking is this, that the eye of the mind be never taken off from things themselves, but receive their images as they truly are. And God forbid that ever we should offer the dreams of fancy for a model of the world.

This was undoubtedly an expression of Bacon's strong aversion to the fanciful theorizing of the Greek philosophers as well as their successors, theorizing which appeared to get nowhere because it was wholly disconnected from experience and never subjected to experimental test. Yet we must now admit that the success of modern science, particularly in the realm of physical science, is due precisely to the willingness of the scientist to "offer the dreams of fancy for a model of the world." In other parts of his treatment Bacon admits that "fruitful natural philosophy has both an ascending and descending scale of parts, leading from experience to axioms, and from axioms to new discoveries." If we can interpret his "axioms" as the hypotheses of theories, his statement would not be far from the view of the nature of scientific theory discussed above. However, he apparently feels that the "axioms" are to be set up from "experience" and hence the imagination should play a minor role at best.

An examination of the history of science, particularly physics, reveals that in the construction of the earlier theories the attempt was indeed apparently made to make the hypotheses mirror experience as directly as possible. This is notably true in the principles of mechanics, where for example the principle of inertia (first "law" of motion) certainly seems to us today suggested by experience in observing the motion of bodies. Torricelli's explanation of the behavior of the barometer tube (tube full of mercury inverted in a dish of mercury) in terms of the atmospheric pressure seems to require only that the notion of the pressure exerted at the bottom by a liquid column be extended to a gas as well. The hypothesis that sound is due to the propagation of motional disturbances in air clings pretty close to experience. Examples could be multiplied indefinitely. However, when we come to contemporary theories we meet a different situation: experience had little to do with establishing the postulates of the quantum theory or the more recent quantum mechanics. Here creative imagination has had free play and the marvelous thing is that it has worked!

Actually, a closer scrutiny of the earlier theories which seem at first glance (to us) to have so direct a grounding in experience will show that even there creative imagination was at work to a greater extent than usually admitted. Certainly the predecessors and even the contemporaries of Galileo did not in general think that the principle of inertia was closely suggested by experience; rather, they could not understand it at all, since they felt that an object could not move without having the moving influence continually exerted on it. In terms of the cultural and scientific atmosphere in which it was first developed, the idea of inertia was a bold act of the imagination and by no means obvious from experience. Certainly it was not an induction in the Baconian sense. Nowadays we use mechanics so freely we tend to take its foundations for granted—in fact, to most engineers it doubtless seems to have no hypothetical character at all. To appreciate the theoretical basis of science it is evidently necessary to have some acquaintance with its historical development.

In any case, modern scientists feel no compunction about introducing any sort of hypothesis as long as it works, and in physics the tendency seems to be in the direction of the more abstract type of postulate bearing little obvious relation to observed experience, though it may be suggested by certain mathematical algorithms.

The realization that scientists make free use of the imagination in the construction of scientific theories has important implications for the relation between science and other forms of coping with human experience. As we shall stress in a later chapter, it means that scientists indulge in preferences and show tastes and prejudices— in other words, they employ *value judgments,* just as other intellectuals do. The clever ones, however, are able to "sell" their "images" and set a fashion for looking at things in a certain light. A particular theory may turn out to be so successful as to persuade many that it is the "final" answer to the problem of understanding a certain domain of experience. But we must not forget that its fundamental basis came out of the imaginative mind of a man or the joint imaginative efforts of a few collaborators. The problem is indeed more complicated than these simple sentences would suggest. We shall return to it in the next chapter.

For the moment, we want to explore one consequence of these considerations. This is the essentially *arbitrary* character of scientific theorizing.

Arbitrariness in Scientific Theorizing

Among the questions raised by the account of scientific theorizing presented in the foregoing sections is the following important one: Does science as a method for the description, creation and understanding of experience proceed inexorably and irresistibly toward a complete and final understanding of all experience? To the believer in the existence of a real world lying outside the observer, this question would mean: Does science guarantee that we shall ultimately know and understand all the properties of this external world? This view equates the task of science to the *discovery* of something objectively "out there," wholly independent of the existence of human observers, and supposes that if we keep on scientizing we shall eventually learn everything there is to learn!

It is not our purpose here to enter upon the philosophical controversy between idealism and realism, with which the above might seem to be related. Actually, it is usually agreed that the successful pursuit of science does not depend on adherence to either of these philosophical points of view. However, granted that science deals with experience without the need for faith in any particular philosophical theory of the ultimate nature of this experience, it yet remains vital to be clear on the issue whether science is more successful if considered as *discovery* of something "there" or as *invention* that can be subject to continual change in the light of the continuous creation of new experience. The reason this is important is simple: if science is discovery, every successful theory tends to become solidified as scientific dogma, i.e., there is a tendency to endeavor to fit all newly obtained experience into its mold under the impression that the final pattern of understanding has been reached. This leads inevitably to inflexibility and, ultimately, as the history of science has shown, to frustration. If, on the other hand, science is invention, the mind of the scientist is left free to explore all possible imaginative avenues in the endeavor to understand new experience, and thus new theories are more freely created, which in turn stimulate the search for still newer experience. It is hard to resist the conclusion that the latter viewpoint is better calculated to advance the whole undertaking.

But there is a price to be paid. This lies in a certain degree of arbitrariness in scientific theorizing. Arbitrariness is here equivalent to the free use of preference in the construction of concepts and theories. This introduces into science an important role for value judgments, commonly considered to have a place only in the humanities. We have already mentioned this in passing in the preceding section and expect to go into the whole matter thoroughly later on. For the present we merely wish to emphasize the essential freedom of the scientist in the building of theories. This is usually overlooked in the current scientific treatises which naturally stress the currently popular points of view and often give the impression of finality. But a historical examination of the literature of science shows how enthusiastically scientists have availed themselves of this freedom to hypothesize, to create new ideas and to attempt to relate them to the experience of their day. We have already called attention to the numerous instances in which alternative theories provided explanations for the same set of phenomena, and the adherence to the one or the other depended largely on prejudices based on allegiance to the scientific prestige of the author and similar adventitious reasons.

We would be wrong, however, to leave the impression that scientists as a class are reconciled to the degree of arbitrariness in theorizing implied in the above remarks. It is only natural for an investigator who has thought long and hard about a certain aspect of experience and has made himself master of the principal facts to feel that once he has got an idea which seems to fit, it is the "right" idea. To him such a notion no longer seems arbitrary, it comes dangerously close to being the "truth." And it is probably a good thing that this is so, for it is unlikely that science would progress if scientists had not an enthusiastic faith in the value and ultimate success of what they were doing. At the same time it is equally good that all theories are subjected to the criticism of those who can approach them more dispassionately than their authors.

What is the value of stressing the essential arbitrariness of scientific theories? It is simply to emphasize the tentativeness of science as a method for understanding experience. Modern science as the inheritance of the Renaissance does not look upon experience as the Greeks for the most part did and does not indeed ask the same kind of questions about the world. If history means anything, the suggestion is strong that our descendants will in their turn ask different questions from ours and develop theories with quite different concepts and hypotheses. This view is of course predicated on the expectation that the domain of experience will grow steadily larger as applied science provides more and more complicated hardware for man to play with. It suggests that science is a kind of game with an infinite number of counters, in which local and momentary successes are possible but in which there is no such thing as a final win. This further intimates that the undertaking is most appropriately carried out in a spirit of humility and not with the arrogance of dogmatic assurance.

CHAPTER 3

SCIENCE AND
THE HUMANITIES

A Preliminary Comparison

In this chapter we explore the relations between science and other forms of coping with human experience. It is natural to begin with the humanities, since so many humanists (to use the term here as referring to scholars in the humanities) still feel that science is a closed book to them, and that there is no relation between what they do and what scientists do; at the same time, many scientists are scornful of what they consider the impracticality of the humanities. They remember, too, perhaps, what a battle had to be waged in the nineteenth century to get science admitted on an equal footing into the educational programs of the British and American institutions of higher education and the rather actively hostile role played in this matter by many humanists.

These bickerings are unfortunate in the face of the growing need for all intellectuals to put on a united front against the anti-intellectual tendencies of the mid twentieth-century scene and the misunderstanding by the lay public of what scholarship means in all fields. There are hopeful signs that better mutual understanding is gradually developing along with a growing appreciation of the similarities between science and the humanities to replace the hitherto all too vocal emphasis on differences.

Of course the differences are obvious to most; the similarities are more subtle. This demands some examination. The communication engineers John Mills and Joseph B. Maxfield of the Bell Telephone Laboratories once made a very interesting observation about the behavior of people when confronted by two objects or situations and asked to compare them. Some and indeed the majority will at once comment on how different they are. The others, the minority, seek to find similarities. Mills called these the D and S people, respectively. By a few simple experiments he thought he could identify them and that the characteristic thus identified has an important bearing on their occupational success, at any rate in the field of engineering. It is unnecessary to go into this vocational side of the matter. We merely note that, as far as science and the humanities are concerned, most observers are D people. Few have shown the S characteristic, and this is perhaps understandable. Mills and Maxfield found that in general S people make the better scientists, while D people make the better business executives.

Before going further it is desirable to identify the humanities as we shall understand them. In institutions of higher education the term is commonly applied to the ancient and modern languages and their associated literatures, philosophy, religious studies, music and art. The position of history is ambiguous. Though usually classed as a social study, it possesses certain characteristics, e.g., the possible literary quality of its presentation, that ally it to the humanities already mentioned. We shall look into this later. Possibly mathematics, in so far as it is a language, belongs with the humanities. This also demands further investigation.

The subjects just mentioned have as their common domain the problems of the individual human being and his response to his environment. Philosophy and religion strive to answer man's most absorbing questions: Who am I? Where did I come from? Where am I going? What ought I to do? In literature, man tries to set down in the symbolism of ordinary language (as distinct from the technical jargon of science and technology) the story of the individual's life on earth, his joys and sorrows, loves and hates, successes and failures; in fact, all those fundamental implications which all of us somehow understand no matter how sketchy and unprofessional our formal education. The basic appeal of literature is to the interest each of us as a human being feels in other human beings, an interest which appears to transcend all others that we display. Poetry as a special form of literature pushes language symbolism to its highest esthetic pitch. Music and art employ symbolism for esthetic purposes in direct appeal to the ear and eye respectively.

The reader at this point will probably wish to raise a question. Who are the people who represent the humanities? Are they the university professors who teach the subjects mentioned or are they the people who have created the stuff which is taught? The answer is, of course, that both groups are essential and both can be called humanists. If history means anything, we must conclude that the number of genuinely creative people in any line of endeavor will always be relatively small. This is a psychological mystery which presumably will be cleared up some day. In the meantime, the impact of creativity on the general mass of people could hardly amount to much were it not for the efforts of scholars who by studying the works of the creative ones manage to explain and interpret them to the masses. In so doing they insure that knowledge of the creations of the great artists, musicians, poets, novelists, etc., shall be preserved for the benefit of the race. Moreover, through the education of the young they guarantee that there shall continue to be an awareness of these creations among persons who wish to think of themselves as civilized. Such service of preservation and education should not be scorned. It is unlikely that the humanities could have survived the buffetings of time without it.

Now that we have defined briefly the humanities and characterized the persons who represent them, we face the question, what are the alleged differences between them and science. We can summarize them as follows. The humanities are said to be subjective in their approach to experience, whereas science is objective. Science is coldly factual while the humanities deal with values. Science is a public affair and makes its impact on all persons independently of their desires or opinions; the humanities are intensely personal and individualistic.

Before analyzing these alleged contrasts it will be well to clarify their meaning.

By objectivity most persons mean the characteristic displayed by the scientist in describing things as they really are without allowing personal prejudices to enter or the emotions to color the observations. The scientist seeks to establish agreement among all normal observers of the domains of experience in which he is interested. He tries to describe that part of experience which is common to all competent observers. The humanist, on the other hand, emphasizes the role of the individual subject in the handling of experience. He is concerned with the way each individual reacts to his environment. Thus, if the humanist is an artist, e.g., a painter, he creates a picture which represents what his surroundings seem like to him (we are at the moment neglecting abstract art); he is not interested in a photographic reproduction. Inevitably he puts a great deal of what we call his personality into the thing he creates. Other persons judge its worth not primarily as a representation of common experience but rather as an expression of the artist's subjective feeling, and, of course, give it their own private interpretation.

In the second place, science is said to be concerned principally with facts, while the humanities in addition involve values. On this view normative statements, e.g., expressions of taste, choice and preference, have no place at all in science, whereas they are a leading preoccupation of the humanities. Science does not tell us how to lead happy lives, it merely describes what life is observed to be like. The humanities discuss what many admittedly wise men have thought and said on the fundamental questions so crucial for the behavior and fate of the individual. The fact that this alleged difference is grossly misleading, as we shall have occasion to point out a little later on, has not prevented its dissemination. It looks plausible to the casual observer who is not too well acquainted with either science or the humanities.

It is continually stressed that science is public whereas the humanities are private. This is closely connected with the objectivity-subjectivity contrast noted previously. Though science is created by individual scientists, it establishes its right to be taken seriously only when it gains acceptance by a large group of qualified persons, or so it is held. There must be agreement as to what is observed and how this is talked about. But the individual humanist can take whatever view he pleases about anything or create anything he likes; it is his feeling and tastes he is expressing and he is not bound to respect those of anyone else. Whenever he makes an impact, so to speak, it is on another individual. Suppose he is an artist and paints a picture. He does not do this for the sake of gaining public approval (or so it is alleged) but to please himself. When someone looks at the picture, he again judges it as an individual who can approve or disapprove as he sees fit. If enough observers approve, perhaps the picture is purchased, hung in a gallery and talked about by critics. Only then does it begin to make an impression on the so-called public. With science it is supposed to be different. Scientific theories lead to laws which can be tested in the laboratory. If confirmed they hold for everyone. No question of choice is involved. One may choose to be ignorant of them, but they may not be ignored safely by anyone. If they lead, as is, eventually, usually the case, to applications in the form of gadgets, these gadgets work for the whole public and they affect the whole public. No one can really escape them whether he is interested in science or not, in the sense in which he can escape contact with music, art or literature if he so chooses (in spite of a so-called liberal education). In conclusion, the impact of science on civilization is universal; the impact of the humanities is selective and individual. If this difference is genuine the social implications are very considerable.

One final point at this place! It is often stated that because science has to describe things as they are, it cannot provide scope for the imagination, whereas the humanities throughout provide an imaginative interpretation of experience. No one can question the latter part of the assertion. Those who have followed closely the developments in Chapter 2 will see reason to doubt the former part, since we have stressed the imaginative aspect of the creation of scientific theories. But we shall go into a detailed comparison on this point further on.

Similarities. Objectivity and Subjectivity.
Public and Private Impacts

The previous section set forth briefly certain alleged differences between science and the humanities. The task now is to see whether these differences are genuine, and whether there are real similarities which balance or even outweigh verified differences. Let us try to abandon the D point of view and be S people for a while!

In the first place, we should dispose of a problem which might have been raised when we were talking about the two kinds of humanists, i.e., those that create and those that talk about the creations and help to preserve them and their meaning for posterity. Does this distinction prevail among scientists? It certainly does! The number of scientists who create theories and do the epoch-making experiments which either suggest or are suggested by them is relatively small. Most scientists spend their time applying the predictions of well-known scientific theories to so-called practical situations, or they work out further consequences of such theories. They may conduct experimental measurements for the more precise evaluation of fundamental constants such as the velocity of light or the acceleration of gravity. They may develop new gadgets based on existing knowledge. A good many scientists teach in colleges and universities and hence, like their colleagues in the humanities, assure that the knowledge of science is passed on to our successors. Even when the teachers do research, as it is understood they are under some obligation to do, it is rarely of fundamental character, though there are certainly exceptions. However, the principal point is that science and the humanities do not differ in the respective classifications of activity. This in itself is a bond of union, an important similarity.

We now pass on to consider more carefully the specific alleged differences mentioned previously. First, there is the matter of the apparent objectivity of science contrasted to the subjectivity of the humanities. This side of science stresses its impersonal character: a description of experience is not regarded as scientific unless it is agreed to by a set of independent, impartial observers, whose personal prejudices are somehow cancelled out in their ability to find the same thing in the phenomena in question. We do not talk, for example, of establishing a scientific law like that of Boyle unless several observers can get the same result after going through the requisite experimental operations. There is no question here, clearly, of taking an emotional attitude toward the experience. One merely agrees to carry out a certain set of operations and describe the results as honestly as possible. When agreement has finally been secured by a sufficiently large number of observers, the objective quality of the experience and its description is then assured and we speak of a scientific law. This is, at any rate, the standard view and it is foolish to deny that it has much validity. Scientists do strive for objectivity and wish to have their ideas verified by their colleagues.

But as we have seen in Chapter 2, this is by no means the whole story. Science does not consist merely in the description of experience in terms of laws or in the creation of experience through experiment. It also strives to understand experience by means of theories.

Now a theory is a creation of the mind of an individual and as such involves intimately his own personal attitude toward the experience he is striving to understand. In other words, it cannot fail to have intensely personal and hence subjective aspects. Naturally, the ingenious creative scientist seeks to obtain the interest and sympathetic hearing of his colleagues, but so presumably does an artist when he creates a work of art or a writer when he pens a literary work. Undoubtedly, the fact that certain theories have proved so successful in subsuming known experience and predicting new experience gains for them a certain objective character which the creations of humanists rarely seem to possess, and yet we know that there gradually grows up agreement among critics as to what is good art and what is poor art. Moreover, as fads come and go in art and literature, so fashions change in scientific theories.

So, in conclusion, it appears that in so far as science depends on theorizing for its understanding of experience, its objectivity has a subjective quality that reminds us of the humanities. At the same time, the alleged subjectivity of the arts in practice is more objective than is commonly admitted.

But it may be objected that a more decisive difference between science and the humanities resides in the obvious public impact of the former as contrasted to the private character of the influence of the latter. This has already been stressed earlier in the book. No one can doubt the profound influence on society of science through its technological applications, and certainly these affect all persons whether they like it or not. A person need have no acquaintance with science in the way we are interpreting it to have his life modified by it through the various gadgets he uses every day. This is public impact with a vengeance! On the other hand, persons who do not choose to look at works of art or read works of literature can pretty well escape their influence. Or so it would seem, until we reflect that the state makes an effort to give every person of normal mentality a school education in the course of which he is exposed to certain of the humanities. Moreover, newspapers, radio, and television, all the so-called mass media of communication, transmit information which has, on the whole, more relevance to the humanities than to science.

So perhaps it is a bit hasty to conclude that the humanities are unlike science in that their influence is on the individual rather than on the mass. Actually, it acts on both. Each individual may well react differently to works of art, music and literature, but if there were not in addition an effective public reaction, theaters could not operate, there would be little purpose in the construction and maintenance of museums, and symphony orchestras would go out of existence.

In the meantime, in this matter of public versus private impact, we have done science an injustice in neglecting its ideological side. The ideas of science in so far as they reflect the imaginative powers of the creative scientist can and do make a powerful individual appeal to many thoughtful persons, who are not themselves excited by the applications of science. It is hard to think of this appeal as being in essence different from that which a work of art or literature makes on a sensitive person; though admittedly the number of people affected at any given time in history may well be smaller.

Similarities. Value Judgments

The problem of the alleged difference between science and the humanities on the score of value judgments deserves a whole section by itself. As we have already seen, the common view is that science deals primarily with facts, while the humanities are heavily weighted in the direction of values and value judgments. Specifically, it is claimed that science has nothing to say about the larger issues of life. It would appear to have no room for the expression of tastes and preferences, since it must always stick to things as they are.

Anyone who has studied and carefully thought through the analysis of the nature of science presented in Chapter 2 will recognize that the allegations of the preceding paragraph are false. For we have already seen that in the attempt to understand experience, scientists construct theories. Theories involve the introduction of hypotheses and this is always the expression of a taste or preference, though the very success of certain scientific theories often promotes the belief that the hypotheses are really forced on the theorist by experience itself. That this is not really the case can be readily seen by some illustrations from the history of science.

We recall first the problem of the nature of light. Many of the phenomena associated with it, such as reflection and refraction, have been known for ages. Other properties like dispersion, diffraction, interference and polarization are of more recent discovery or detailed investigation (the seventeenth through the nineteenth centuries). If these experimental phenomena and the associated laws had been by themselves sufficient to dictate rather unambiguously a theory of light and its fundamental hypotheses, there presumably should be only one successful theory. As a matter of fact, there have been several. Two of these have been of primary importance in the historical development and are indeed so even today. They represent two differing tastes or preferences as to the fundamental nature of light.

It is well known that Sir Isaac Newton preferred to look upon light as due to the motion of tiny particles or corpuscles, themselves invisible but able to produce the sensation of vision on striking the retina of the eye. This was the so-called corpuscular theory of light which dominated physics during the seventeenth and eighteenth centuries. It did not of course go unchallenged, since Christian Huygens preferred to look upon light as a wave phenomenon similar to sound. He championed the wave theory of light and was able to show that this fundamental hypothesis accounts satisfactorily for the experimental laws of reflection and refraction as well as the more elaborate phenomena of double refraction and polarization.

The choice of hypotheses in the theory of light cannot be regarded otherwise than as an expression of value judgment. One thoughtful individual in full possession of the experimental facts preferred to take one point of view and undoubtedly felt it was right. Another equally competent scientist could not accept this point of view and preferred a quite different one. Neither point of view was forced on its adherent—it was a free choice so far as any modern interpretation can tell. Naturally Huygens may have felt himself "forced" to adopt the wave hypothesis, but it was clearly only a personal matter, since otherwise how could a distinguished investigator like Newton have felt so differently? Each without doubt felt that time would prove his view correct. Is this situation essentially different from the expression and maintenance of value judgments by scholars in the humanities? One scholar or writer prefers realism in literature, another values romanticism more highly. These are judgments which mean much to the individuals concerned and to others who are swayed by their persuasiveness. But they are in the last analysis arbitrary and there is no way to establish anything like essential correctness about them. It is precisely the same with scientific theorizing.

We shall return to the theory of light a little later, since the observant reader has doubtless already thought of some critical questions he wishes to raise. First, however, let us look at another example of the existence and importance of value judgments in science. One of the most absorbing puzzles of science has been the nature of matter, i.e., what things are really made of. From the earliest recorded history it is clear that speculation on this subject has been universal. One of the earliest theories to emerge, in vague enough fashion to be sure, was the atomic theory, which says that all matter is ultimately made up of collections of tiny particles which themselves are not further divisible. This assumption is a value judgment. Those who make it are not in a position to demonstrate its truth objectively. They are attracted to it by reasons which probably must be classed as psychological if we wish to pretend to be scientific in the assessment. Others who for various reasons do not like the concept (Aristotle in ancient times) prefer some other view. One might suppose that one view would ultimately prevail, since the evidence supporting it would become overwhelming. Certainly this was the feeling of many distinguished scientists in the late nineteenth century, e.g., men like J. Clerk Maxwell and Ludwig Boltzmann. But there were other well-known physical scientists who did not at all agree with this point of view. Ernst Mach, the celebrated Viennese physicist and philosopher of science, had what might be termed a genuine hatred of the atomic theory and he was joined by the German physical chemist Wilhelm Ostwald. Both these gentlemen and especially Mach developed what later came to be called a positivistic attitude toward scientific concepts which deprecated the introduction of imaginative notions which were not directly suggested by experience. They believed that all physical phenomena which were describable by the atomic theory could equally well be explained in terms of the concept of energy, the fruit of the earlier development of mechanics and thermodynamics. This concept demands no particular special assumption about the constitution of matter, and might therefore be deemed a more economical way of looking at things than would be the atomic theory, which in any case also incorporated the notion of energy in its structure. Mach became the chief late-nineteenth-century exponent of the belief that one of the chief virtues of science as a means of handling experience lies in its so-called "economy of thought." From this standpoint, his negative attitude toward atoms and his adherence to energetics, as the other point of view came to be termed, were not surprising.

The main thing for our present purpose is simply that Mach and Ostwald were expressing value judgments when they preferred the energetical to the atomic method of scientific description. On the other hand, Boltzmann, one of the creators of statistical mechanics and the statistical method for studying the properties of matter, made a different value judgment in staking his scientific standing on the atomic idea. Moreover, we must not suppose that these attitudes were casually and tentatively held. These men were serious and felt rather dogmatically that they were right and those who thought differently from them were wrong. They were willing to stand up and argue heatedly for their views. It is reported that on one occasion when Boltzmann was lecturing, Mach was in the audience. A specific reference to atoms by the lecturer was interrupted by Mach, who apparently could stand no more and, rising from his seat, addressed the speaker with the crucial question: "Have you ever seen an atom?" Boltzmann's reply has not been preserved, but in his popular essays he has some hard things to say about the energeticists. He also took his value judgments seriously. One hardly knows whether his ultimate suicide may not have been due to his taking them too seriously.

But surely, the reader well acquainted with the history of physics will exclaim, we now know beyond the shadow of a doubt that atoms exist, that Boltzmann was right and Mach was wrong, and the problem is no longer one of values, but merely of finding out the facts by patient searching! This is a tempting view but it will not stand too close scrutiny. Before we give it this scrutiny, let us look at some other examples, old and relatively new, of choices and preferences in science.

The theory of mechanics is one of the most firmly established of physical theories, so solidly established indeed that most people who use it extensively, like engineers for example, no longer think of it as a theory at all. One might naturally suppose that there is nothing left in it to argue about, that everyone agrees about all its concepts and hypotheses. This, however, is by no means the case. The number of different points of view about the foundations of mechanics is very large. Most elementary texts in physics stress the Newtonian formulation in terms of the famous three laws of motion. This effectively sees change in motion as due to forces, and the problem involved in any particular motion, e.g., planetary motion, as one of finding the correct force to use, e.g., the inverse square central variety. One might term the Newtonian form a causal approach to motion with force as the cause of acceleration, etc. But there are logical difficulties which began to bother keen-minded physicists in the latter part of the nineteenth century, with the result that numerous reformulations of mechanics were undertaken. As a matter of fact, some of these came much earlier. P. L. M. Maupertuis, J. L. Lagrange and W. R. Hamilton took a different view of the foundations of mechanics. They thought of motion as due to the effort of Nature or of God to make certain integrals of quantities having the dimensions of momentum or energy a minimum subject to certain conditions. This view obviously has purposive or teleological implications. Once again value judgments are involved. No one questions the validity of the laws of mechanics which describe actually observed motions of particles or bodies. But quite different preferences arise as to how the fundamental theory shall be organized and what types of hypotheses shall be employed.[1] Even though the various hypothetical approaches lead to the same results as far as obvious experience is concerned, physicists differ as to the value they place on them and they set great store by these value judgments since their subsequent theorizing may be greatly affected by their choices. Consequently, these judgments are by no means a matter of indifference; they may, indeed, have a great deal to do with the future development of science.

The illustrations of value judgments in science have been chosen exclusively from physical science. This limitation is not necessary. There are many examples in other branches. Consider for example the controversy between the mechanists and the vitalists in biology,

[1] See R. B. Lindsay and H. Margenau, *Foundations of Physics* (New York: Dover Publications, 1957), Chapter 3.

or that between the direct creationists and the evolutionists in the problem of the development of living organisms. In geology, some chose the catastrophic explanation for the formation of the earth's crust, others the uniformitarian view. Everything we have said about physical hypotheses above can be equally well stressed about these biological and geological theories.

The reader may now be willing to admit that a case has been made for the existence of value judgments in science, and yet he may be somewhat suspicious of the relevance of the illustrations since they refer to controversies in the past. Rather impatiently, he may suggest that this is now an old story and modern science has got over it, that science has developed so successfully that in contemporary circles everyone agrees about the basic point of view, and hence if there are value judgments, in the sense discussed above, they are now meaningless and irrelevant, since every scientist's judgment is like that of every other one, and the happy, interesting and fruitful differences of opinion prevailing among humanists in their attitudes toward experience no longer exist among scientists. The actual facts render this position untenable. We need present only one example, chosen from modern physics, but there are many equally strong ones throughout all branches of contemporary science.

It is well known that contemporary atomic physics is described by the theory called quantum mechanics, a somewhat formidable mathematical structure which will not be presented here.[2] We should like, however, to call attention to the fact that there has existed from the 1920's, when the theory was first constructed on the foundations so ably laid by Niels Bohr in 1912, a fundamental difference of opinion concerning the physical interpretation of the mathematical formalism. One group has held (and this view is predominant at the present time) that no single atomic event, such as the change in energy of an atom with the absorption or emission of radiation, can ever be considered as determined precisely from pre-existing conditions. The best the theory can do is to predict the probability of the occurrence of any given change and from such probabilities to calculate average values of the fundamental quantities measured in atomic experiments, e.g., momentum and energy. It can indeed predict ex-

[2] *Ibid.*, Chapter 9.

actly what values of such quantities are *possible* for a given atomic system, i.e., the so-called eigenvalues, but it is not able to predict *which* one will show up in any given measurement. In brief, from this point of view, quantum physics is fundamentally indeterministic.

Now there are several physicists to whom the notion of determinism in physics has rather significant value and who are reluctant to subscribe to the indeterministic interpretation of the Copenhagen school (so-called because Bohr has lent the weight of his prestige to this point of view and has indeed supported it with numerous ingenious arguments). Among these are E. Schrödinger, one of the founders of quantum mechanics through his development of wave mechanics, and even to a certain extent (though his attitude is not completely clear) L. de Broglie, another of the famous pioneers in this field. More recently, D. Bohm and J. P. Vigier have come out unequivocally in opposition to the indeterministic view and have endeavored to devise a new interpretation in consonance with determinism. The effort has given rise to much controversy into which it is unnecessary for us to go here.[3] The main point is simply that even in contemporary science, value judgments are being made and defended as important elements in the structure of science. Values enter into present-day science as they have throughout the historical evolution of the scientific method.

We might well rest our case at this point. But those who have been accustomed to think of science as completely objective, positivistic and factual, and hence quite different after all from the humanities with their prime emphasis on values, perhaps still have their nagging doubts. Suddenly the recollection of the term "crucial experiment" comes to mind and they inquire with some satisfaction: Does not this provide precisely the clue to the problem of apparent values in science, for cannot the crucial experiment decide among conflicting theories and by picking out the one that really works leave the adherent of the others in an untenable position, no matter what their value judgments may be or how much they are attached to them?

---

[3] See *Observation and Interpretation,* the Proceedings of the Colston Symposium at the University of Bristol in April, 1957. This contains articles by Bohm, Vigier, Rosenfeld and Popper on both sides of the controversy.

The notion of a "crucial experiment" is a particularly attractive one. Consider the famous example in the theory of light. It will be recalled that both the corpuscular theory in the form advocated by Newton and the wave theory of Huygens predict the existence of the laws of reflection and refraction of light at interfaces between different media as verified experimentally. However, in order to predict Snell's law in its proper form, i.e., make the index of refraction have the proper relation to unity in terms of the properties of the media, it is necessary in Newton's theory to assume that the velocity of light for a given color is greater in a dense medium (in which the refracted "ray" is bent toward the normal) than in the rarer medium from which it comes. On the other hand, the wave theory of Huygens to accomplish the same end has to assume that the velocity of light is less in the dense than in the rarer medium. Here then is an apparently clear-cut difference, and one that seemingly should be readily settled by experiment. Up to the middle of the nineteenth century, the experiment had not yet been performed since no one had yet devised an adequate terrestrial method for measuring the velocity of light. Finally, Fizeau did this with his famous toothed wheel technique. At about the same time, Foucault introduced the somewhat similar rotating mirror method, and both men using the latter scheme for measuring the velocity of light in water in 1850 showed conclusively that it is less than the value for the corresponding color in air. This would appear to settle the matter. The corpuscular theory fails the crucial test, the wave theory passes it and hence the scientist no matter how strongly attached he may be value-wise to the corpuscular viewpoint feels he must discard it.

The search for and performance of crucial experiments then seem to provide a valid method for disposing of the value-judgment problem in science. But let us look at the matter more soberly. Does the measurement of the velocity of light in water really provide a crucial distinction between the wave and corpuscular theories of light? The answer is no. What it does do is to distinguish between Newton's version of the corpuscular hypothesis and the wave hypothesis of Huygens. As Pierre Duhem has emphasized,[4] there is really no way to decide from the experiment in question just what feature of the

[4] See his *The Aim and Structure of Physical Theory* (Princeton University Press, 1954), Chapter 6.

Newtonian formulation ultimately produces the result which is in contradiction with experiment. In fact, it is perfectly possible to construct a corpuscular theory of light which will give the same results, as far as reflection and refraction are concerned, as the wave theory. In any case, we recall that after being banished from the theory of light for half a century, corpuscles came back into the theory with the advent of the quantum theory, and today we have no hesitation in talking about photons or light particles and use them freely in discoursing about the photoelectric effect and other optical phenomena.

Duhem makes out a good case for the conclusion that there is really no such thing as a crucial experiment. At any rate, it does not appear likely that values in science to which scientists are strongly attached simply because, so to speak, they "feel them in their bones," are going to be dismissed lightly by any single so-called crucial experiment. Naturally, if a view like the mechanical theory of heat, for example, seems to fit all the experimental facts more plausibly than an opposing view like the caloric theory, there is not much question where preference will lie, but the important point is that it will still be preference. A similar situation prevails with respect to the famous controversy over the value of the atomic hypothesis and the problem of the "existence" of atoms. Few, if any, contemporary physicists question the atomic theory as an extremely valuable mode of understanding the properties of matter. In fact, scarcely anyone doubts the "existence" of atoms and their various constituent parts, the many so-called elementary particles. It is commonly said that experiments like the diffraction of X-rays by crystals and the production of tracks in cloud chambers actually exhibit the positions and paths of such particles in such categorical fashion that their existence is just as real as that of large-scale physical objects like stones and tables. But it remains a fact that what we observe in all such experiments are marks on photographic plates or small droplets of water, and we must still decide on the interpretation to give the observations. When we decide that the "best" interpretation is in terms of the atomic hypothesis, we are still speaking with a value judgment, no matter how successful the theory may turn out to be in subsuming a mass of data and predicting new experience.

Our chief aim in this section has been to stress a fundamental similarity between the humanities and science in the existence in both disciplines of value judgments. No one doubts the importance of the role they play in the humanities, in the decisions, for example, as to what constitutes good music, good art, great writing, etc. But they play a similarly important part in the progress of science. The maintenance of a preference for a certain theoretical point of view on the part of an admittedly distinguished scientist can exercise enormous influence on the development of a certain domain. Thus, Newton's advocacy of the corpuscular theory of light colored the development of optics for the better part of a century. It is just here that the existence of value judgments in science can lead to a certain danger. For dogmatism is not confined to theology; it can also be exhibited by scientists and can hamper progress. Only the free competition of imaginative ideas is compatible with genuine scientific progress.

Further philosophical as well as physical implications of value judgments in science will be postponed to subsequent chapters.

Special Relations. Music and Acoustics

The general discussion of the relation between science and the humanities in the preceding sections suggests the worthwhileness of a somewhat more detailed account of the mutual interaction of special fields. This will illustrate some of the earlier generalizations.

We begin with acoustics and music. Acoustics is the science of sound, though the name comes from the Greek meaning hearing. Music, of course, is the art of producing sounds which the ear finds agreeable or can learn to find agreeable. Its origin as a human activity is lost in the mists of antiquity. It is safe to say that no major civilization has ever existed without an interest in music.

People made musical sounds long before they were interested in studying them scientifically. Yet it was through interest in music that some of the first notable discoveries in acoustics were made. In the Greek world one thinks at once of Pythagoras (ca. 500 B.C.), who is said to have been impressed by the fact that of two stretched strings fastened at the ends the shorter one emits a note of higher pitch when plucked. In fact, he is supposed further to have observed that the string of half the length emits a note an octave above the other. Since Pythagoras left no writings, the story is legendary. It is usually cited to provide a basis for the zeal that Pythagoras and his followers displayed for integers as basic for the understanding of experience. It seems clear that the germ of the idea that pitch depends somehow on the frequency of vibration of the sound-producing object was in the minds of Greek philosophers of the Pythagorean school, such as Archytas of Tarentum, flourishing around 375 B.C. A fairly clear presentation of this view is to be found in the writings on music of the Roman philosopher Boethius in the sixth century A.D. For the detailed experimental basis, we presumably must look to Galileo, in whose famous *Discourses Concerning Two New Sciences* (1638) there occurs a definite statement that higher pitch is associated with higher frequency of vibration. For our present purpose, the important point is that the existence of musical instruments and the interest in musical sounds stimulated a desire to understand the nature of the sounds. In this as in many other cases science followed art.

The history of music shows numerous illustrations of the same sort of influence. In fact, it was only natural for the early scientific investigators in the field of sound to use musical instruments as sources since they were recognized as producers of more regular or purer sounds than other sources. Thus the phenomena associated with beats, an extremely important concept in the theory of sound propagation, were studied by Joseph Sauveur, the man who first suggested the name acoustics for the science of sound, by means of organ pipes of different length. Interest in the sounds emitted by vibrating strings led to some elaborate mathematical investigations of the dynamics of the vibrations, beginning with the fundamental work of Brook Taylor (the celebrated author of the famous expansion theorem in infinite series), and going on down through the more detailed mathematical analyses of Daniel Bernoulli, D'Alembert and L. Euler. Much of the modern mathematical theory of partial differential equations of the second order developed out of the interest displayed in the vibrations of producers of musical sound. Similarly, the musical interest in organ pipes led to detailed mathematical study of the propagation of elastic waves in confined columns of air. This in turn stimulated investigation of the absorption of sound in air and other gases, a very important branch of modern physical acoustics.

The performance of music for the benefit of large groups of people early demanded the construction of concert halls. This was usually carried out with due fidelity to esthetic and technological principles of architecture but with scant consideration of the acoustical requirements of good hearing. The results were often bad and even when good, the reasons were not understood. However, eventually the problem was tackled scientifically around 1900 by W. C. Sabine of Harvard, who made the first successful attempt to establish the basic principles of architectual acoustics. Much work has been done since and now it is perfectly possible to lay down in advance of construction the recipe for good acoustics in a concert hall as well as to assimilate the treatment into the artistic lines of the interior.

This brief survey scarcely does justice to the influence which music as an art has had on the development of the science of acoustics. To go more thoroughly into this we should have to recall the way in which interest in singing has promoted the study of the production of voice sounds. This is, of course, closely connected with the hearing mechanism. So it is no exaggeration to see the influence of music on the development of what has come to be called psychological and physiological acoustics.

Is there reciprocity in this matter? In other words, if music has stimulated the advance of acoustics as a science, is it true that the science has had an effect on the art? The answer is decidedly yes, especially in the twentieth century. Modern acoustics has for one thing revolutionized the dissemination of music through radio and sound-reproduction devices like record players. It may be argued whether this benefits music as a creative art, but it must be admitted that it has enormously increased the appreciation of both classical and modern music in the population at large. This is patently an influence of acoustics on the arts.

It is well known that the introduction of electronic sources of sound like the piezoelectric and magnetostrictive transducers has greatly broadened the range of possible frequencies and intensities. Whereas many classical musical instruments are restricted effectively to discrete frequencies over a limited range, modern sources can produce a continuous spectrum of frequencies. This should provide musicians with much greater scope for the exercise of their imaginations. No judgment of the esthetic quality of the results is here intended, but the fact cannot fail to have an impact on the future of music. Already musical composition of the conventional sort is being supplemented by direct recording on tape. This may well alter the symbolic writing of music. Naturally, this is speculative, but so is all future application of science. We could go on to suggest that the acoustical studies now underway by psychologists and physiologists on the nature of hearing will lead to new methods of training the ear to listen to music. At the same time the more systematic use of electronic devices by which singers can actually see on a scope the analysis of the tones they are producing as compared with those they are attempting to produce will lead to better schemes of voice training.[5]

It is probably not too fanciful to anticipate that as a result of better understanding of communication theory (to be treated in more detail later on), in which acoustics plays an important role, esthetic appreciation of music by larger numbers of people may be increased. For example, we know that the environment and general noise level are intimately related to articulation of spoken sound. The connection with music has not yet been carefully studied but would appear to be obvious in a general way.

The Graphic Arts

Since we found music as an art closely related to the science of acoustics, it will not surprise us to see many connections between science and the graphic arts, i.e., drawing and painting and the general representation of natural or other objects on flat surfaces. The study of prehistoric man indicates that he depicted scenes from his environment, e.g., animals, on stone surfaces such as the interior of caves. No one knows precisely why he did this, though anthropological research on present-day primitive aborigines suggests that his motives, like those of modern artists, were mixed. To a certain extent, he undoubtedly made pictures to please himself, for the fun of

[5] For suggestions about this and other possible influences of the science of acoustics on music, see John Redfield, *Music—A Science and an Art* (New York: Tudor Publishing Co., 1926).

it. But it is also likely that these pictures played some part in ancient man's practical relations with the objects depicted. Thus, a drawing of an animal might be connected with his desire to slay the animal for food or to be protected from attacks by it. In this case, the depiction presumably had to be as accurate as possible, and art became a kind of hand maiden to the precise description of natural objects, thus a precursor of science. There is, then, no need to labor the point that painting and drawing as representations of nature have had a close relation to the development of science.

Without a grasp of perspective the painter cannot hope to create on a flat surface the illusion of distance, and hence cannot depict things realistically. But perspective is a branch of the science of optics treated geometrically. Again, without a knowledge of the behavior of pigments the artist cannot hope to utilize color successfully in his painting. But the nature of pigments is a problem in both chemistry and light. To be sure, the knowledge of the painter about these matters does not have to be scientific in the modern sense in order that he may go about his business and produce esthetically pleasing and experientially faithful results, any more than the musician needs to know all about the physical theory of the vibration of strings in order to construct a harp and play on it. In the beginning, art creates its own technology. But ultimately the scientist and technologist who study the tools of the artist cannot help turning up useful techniques to make the artist's task more efficient. For example, an increase in knowledge of the absorption of light by chemical compounds is bound to improve the quality of pigments.

But this is by no means the whole story on the relation between science and painting. We might do worse here than meditate on some of the words of Leonardo da Vinci, one of the great painters of the Renaissance. On painting he said:

If you despise painting, which is the sole imitator of all the visible works of nature, it is certain that you will be despising a subtle invention which with philosophical and ingenious speculation takes as its theme all the various kinds of forms, airs and scenes, plants, animals, grasses and flowers, which are surrounded by light and shade. And this truly is a science and the true-born daughter of nature, since painting is the offspring of nature. But in order to speak more correctly we may call it the grandchild of nature; for all visible things derive their existence from nature, and from these same things is born painting. So therefore we may justly speak of it as the grandchild of nature and as related to God himself.[6]

This may be discounted as a lyrical statement, but it does seem to represent fairly the viewpoint of a great genius who, though he may not have been as much of a scientist as some have claimed,[7] certainly read widely in the scientific literature of his time and understood what he read. His statement puts the relation of science and graphic arts on a more fundamental plane, philosophically speaking, and harks back to the general considerations of our previous sections. Here painting as a method for the depiction of nature partakes of the character of science itself. Of course, this takes no account of that painting which is symbolic of the pure imagination of the artist and is seemingly unconnected with the outside world.

Let us revert for one moment to the technological influence of science on graphic arts. One of the most important adjuncts to the painter is the use of color. The phenomena of color have been extensively studied by optical physicists and a relatively enormous amount of technical knowledge is now available. The use of color in the arts is mainly through the agency of dyes and pigments. A dye is a substance which forms a transparent colored solution in a solvent like water or oil. The color of such a solution depends on the fact that when white light falls on it, part is absorbed and part is transmitted. The color we see is the transmitted part. A pigment on the other hand is a finely divided colored solid (a collection of a large number of tiny particles) usually suspended in some liquid medium like an oil. Like a dye it absorbs part of the white light falling on it, but unlike a dye its apparent color is due to the fact that it diffusely reflects what it does not absorb. It is always used in the arts by being applied with a brush or spray on a solid surface. Different colors can be obtained by proper mixing. The results depend on some important physical properties of the light-absorbing powers of substances. Thus, a common so-called yellow pigment will absorb all but the yellow

[6] Edward MacCurdy, *The Notebooks of Leonardo da Vinci* (New York: Garden City Publishing Company, 1941), p. 859.

[7] See the negative opinion on this point expressed by J. H. Randall, Jr., in his article "The Place of Leonardo da Vinci in the Emergence of Modern Science," *Roots of Scientific Thought*, edited by P. Wiener and A. Noland (New York: Basic Books, 1957), p. 207.

part of the visible light spectrum (the band of colors formed when white light passes through a prism), and some of the green which is adjacent to it. A common blue pigment will absorb all but the blue and some of the green which is adjacent to it. Hence when the yellow and blue pigments are mixed only the green fails to be completely absorbed and the resulting mixture looks green. Of course, the nature of the green will depend on the amount of it not absorbed by the yellow and blue pigments, respectively.

Both dyes and pigments produce color effects by the absorption of light or what might be called subtractive processes. But it is also possible to produce colors by additive processes. For example, when one superposes a certain kind of yellow light on a certain kind of blue light, the addition produces a gray or white. Similarly, red light of a certain wave length when added to green light of suitable wave length produces yellow. Advantage of this kind of effect was taken by the nineteenth-century French painter Georges Seurat, who introduced the style of painting known as pointillism, in which instead of continuous strokes of pigment, dots of different colors are employed. The summation of the effects of the dots at the eye of the observer produces an overall impression both of pattern and color. The scanning techniques used in television are somewhat similar in character, though here we are perhaps outside the realm of art!

From the point of view of the painter, one of the most important characteristics of a pigment is its permanence, that is, its ability to remain unchanged in its physical environment. Light, dampness and foreign gas fumes in the air, all can have a deleterious action on many pigments and it therefore becomes necessary to study these effects scientifically to provide the artist with the best and most lasting materials, as well, of course, as adequate preservatives. Thus reds formed of oxides of iron are more permanent than organic reds like carmine (an extract from the cochineal insect). In other words, dull red pigments tend to be more permanent than bright red ones. Cobalt blue and Prussian blue are almost completely unaffected by light. Various preservatives and protective coatings have been devised to prevent undue aging of colors in paintings and much scientific investigation has gone into these.

In a certain sense it is a pity that the painter has to rely for his color effects on so indirect a vehicle as pigments, when so many more

brilliant colors are provided directly by rather simple physical phenomena. The dispersion of light which produces the colors of the
rainbow and those seen in prisms, the diffraction and scattering of
light which are responsible for the brilliant colors of the sunset, the
interference of light to which are due the beautiful colors of soap and
thin oil films are all, so to speak, natural in the truest sense. The iri-
descence of the peacock's feathers is another illustration of interference. It would be a great contribution of science to art if a way could
be found to make interference colors directly available to the artist,
and it would not be surprising if this were ultimately to take place.

Another aspect of the value of technology to art is found in the
various tests available to ascertain the age and authenticity of paintings. Such aids as X-rays and infrared and ultraviolet radiation are
now commonly used for these purposes. They depend for their effectiveness on the selectivity of absorption of various pigments for light
of various wave lengths. Such nondestructive testing is a great boon
to the art historian and expert.[8]

We should not overlook in this brief discussion of the influence of
science on art the great interest which physiologists and psychologists
have displayed in the nature of art. To investigate what the painter is
really doing as compared with what he thinks he is doing is certainly
a relevant scientific activity and it may conceivably have much to do
with the future of art. Thus, for example, there is the question:
Should painting mirror experience or should it attempt to transcend
experience? If it does the latter, what sort of effect will it have on the
beholder? These in the last analysis are psychological questions, and
critics and philosophers of art will not be able to ignore them. It may
be objected that the artist himself will go his own sweet way without
paying attention to such considerations, but there is evidence that he
is not immune from the effects of the appraisal of his work by others.
These matters have recently been considered by Sir Russel Brain[9]
and R. G. Collingwood.[10]

No student of the psychology and physiology of art can afford to

[8] For details on this and other influences of science and technology, see A. E.
Werner's article "Scientific Techniques in Art and Archeology," *Nature, 186*, 674
(1960).
[9] *The Nature of Experience* (London: Oxford University Press, 1959). See particularly Chapter III on "Symbol and Image."
[10] *The Principles of Art* (London: Oxford University Press, 1938).

ignore the classical contribution made to the relation between physiological optics, i.e., the study of light as perceived by the eye, and painting by the great German physiologist and physicist Hermann von Helmholtz.[11]

So far in this section we have concentrated primary attention on the influence of science and technology on the graphic arts and have said little about the reciprocal influence of art on the development of science. On the technological side, this influence is so obvious as to need little comment. The invention of any device inevitably stimulates some one sooner or later to endeavor to understand how it works and thus promotes science. Historically, much of what we consider science was preceded by rather elaborate technology. In so far as painting is a technical achievement the same situation prevails here. Both the painter and his critics want to know how he manages to achieve his effects of light and shade, perspective and color harmony. Ultimately, a better understanding of the nature of light and the perception of light by the eye results. It is not surprising that many famous craftsmen of the Renaissance turned ultimately to a more detailed scientific examination of the basis of their artistic accomplishments. Thus the famous Italian architect, painter and organist Leon Battista Alberti is believed to be the first to study scientifically the principles of perspective. Somewhat later, the German painter Albrecht Dürer invented etching and became an expert engraver. We do not think of these men as scientists. Yet their artistic work led to great technological inventions and better scientific understanding. They were individuals who not only created beautiful things but also meditated on the method by which the creation was made possible and the fundamental meaning of what had been done. This was notably the case with Leonardo da Vinci.

In their interesting anthology *Roots of Scientific Thought*, to which reference was made above, the editors Philip Wiener and Aaron Noland speculate in their introduction that one reason why the Greeks in their mathematical thinking paid more attention to geometry than algebra may be that their chief arts were architectural, sculptural and pictorial. This is a suggestive idea with a definite relevance to the reciprocal relation of art and science.

[11] *Popular Lectures on Scientific Subjects*, Second Series (New York: D. Appleton and Co., 1881), Lecture III, pp. 73-138.

Literature

One of the greatest of the humanities as a method for describing
and understanding human experience is the employment of language
as a means of communicating thoughts and feelings. When the
thoughts and feelings relate rather directly to human life and are
themselves capable of arousing emotion in addition to conveying
factual information, we may call the result literature. This may take
the form of short stories and novels descriptive of life in its various
aspects, of essays commenting more abstractly on life's problems, of
certain biographical and historical studies, of drama and finally of
poetry. In all these vehicles the aim is not merely to say something
which someone else will find interesting to read, but to say it in such
an effective way that its esthetic value will be recognized by the sensitive reader.

It is hardly to our purpose here to get involved in the argument
whether there is some mysterious *absolute* esthetic value associated
with *real* literature as contrasted to mere writing that transmits information. This is a problem which has presumably puzzled many an
honest young soul when first confronted with works of literature admitted by society to be great. Our stand here will have to be the same
pragmatic one we have taken with respect to the other humanities
and to science itself, namely, that some works of man make such an
impression on enough people that they become part of the intellectual and cultural heritage of the race. It is one of the functions of
learned men to explain why. On the whole, both in the humanities
and in science, they have done a pretty good job in their separate
fashions.

Our problem now is to examine briefly the relation between literature and science. We shall not pretend to do more than touch the
high points; the subject has been exhaustively treated in many quarters, though perhaps too rarely by those who understand both sides
equally well and show equal sympathy with both. Much of the discussion of the first three sections of this chapter on similarities and differences between the humanities and science has dealt so directly
with literature as perhaps the commonly accepted chief representative of the humanities, that we need not go over this ground again.

But there remain some interesting questions. In the first place, has there been a direct influence of science on literature, and if so, what has been its character?

The question is to a certain degree rhetorical. The answer is certainly the same affirmative one as in the case of the arts and music. Let us first confine ourselves to technology. The invention of printing from movable type was undoubtedly one of the greatest boons ever granted to literature. It enabled the writing of a single individual to meet the eyes and impress the minds of thousands and ultimately millions of people and disseminate ideas which had hitherto been the possession of a relatively small fraction of the population. The interest thus aroused could hardly fail to be favorable to the production of further writing of very diverse character as well as to its acceptance by greatly increasing numbers of people. The contribution to culture is obvious. We might indeed have stressed the similar situation in music and the graphic arts: whatever leads to the greater dissemination of humanistic creations ultimately profits both the producer and the receiver. Those who are dubious about the value of the contemporary so-called mass media as disseminators of culture will doubtless challenge this assertion. We refrain from arguing the point here, save to express the opinion that the challenge is probably irrelevant. However, we shall return to this problem later in the book. (See Chapter 9.)

The reader will readily think of other examples of the influence of technology on literature. Such a simple matter as the invention and manufacture of spectacles has been of profound significance, since in order to appreciate literature, people have to read it, and vast numbers who wish to read have defective vision. The invention of the idea of a library as a place for the storage of books and other printed material in orderly fashion calculated to guarantee their use and long preservation was and continues to be a factor in the development of literature and its influence on society. Experiments continue to be made in the search for cheaper and more effective ways to prepare reading matter, e.g., photographic processes and the like in place of letterpress. Experiments in publishing, such as the paperback book, are not too trivial in their impact to be neglected as technological factors in the furtherance of literature and its meaning in our culture.

Since literature uses language as its vehicle the study of language becomes an important element in the creation and criticism of literature. But the study of language is a science. We shall go into this to a certain extent in Chapter 6 and therefore at this point limit ourselves to pointing out that in contemporary information theory language is being subjected to mathematical scrutiny as never before and the ultimate effect of this on writing is bound to be considerable.

So much for technology, though more might easily be said! What has been said will very likely prove sufficiently irritating to many humanists, since it seems to relate primarily to the mechanical appurtenances of literature and not to its essence. Obviously, literary works of great genius were produced long before printing was invented, and the fact that a piece of literature comes to the attention of hordes of people ("best seller") does not thereby make it great literature. One can hear the humanist remark with some scorn: "Will the mathematical study of language provide a recipe for the production of literary masterpieces? I doubt it!" These are arguable points to which for the moment we merely call attention for the meditation of those who feel they are worthy of meditation.

But we must now tackle the much more important problem of the ideological relations between science and literature. A specific example will best initiate the discussion. Jonathan Swift's *Gulliver's Travels* is rarely read in its entirety in these days. This is a pity, since the story of the third voyage, namely, that to Laputa, is a very revealing satire on the science of eighteenth-century England, and in particular the scientific endeavors of the Royal Society of London. Those who have read it will recall the "projector" of the academy of Lagado who was engaged in the attempt "to extract sunbeams out of cucumbers," and the other who had written a treatise on "the malleability of fire." On the more "practical" side was the project for using hogs as inexpensive plows by burying chestnuts, etc., in the ground and allowing the animals to root for them. Swift undoubtedly felt that much of the activity of the fellows of the Royal Society and the research sponsored by them was of the same "useless" sort he depicted in his tale. Whether he really understood enough about science to make his judgment worth anything is open to doubt, but the fact remains that he was commenting on science as he saw it in his time, and hence that very science had an influence on his writing:

there would have been no point to his satire had it not been for the existence of the Royal Society and its avowed purpose to devote itself to the "pursuit of experimental knowledge." The fellows of the Society cast their net pretty wide and came up occasionally with some strange fish, so perhaps they deserved satire. But they represented to the full the spirit of unbounded curiosity without which there can be no science. Even Swift thought it worthwhile to make the behavior of the floating island of Laputa depend on the manipulation of a huge magnet. Though his account of the control of the motion of the island by means of the magnet (or loadstone, as he calls it) is more like crude science fiction than serious science, he evidently had heard of magnetic attraction and repulsion. Unfortunately, he shows no knowledge of the elementary concepts of motion or equilibrium in the Newtonian sense. In this he was to a certain extent typical of English literary figures of his own and later times who have used scientific allusions in their imaginative writings. The far too common procedure has been to fall back on certain well-known facts with a complete ignoring of fundamental ideas. One could hardly in all fairness expect otherwise, since in the case of Newton and his successors in the field of celestial motions, for example, the really important ideas were expressed in mathematical language which in general repels persons whose interests are primarily literary. Why this should be so is one of the mysteries of the human personality, for mathematics is itself a language of great power and beauty.

But leaving aside the severity of the mathematical form of the Newtonian contribution to mechanics and cosmology, there remains the fact that many of the new concepts in science were difficult to grasp by people who had been brought up on what we now think of as decidedly more naïve notions. It took a long time, for example, before there emerged a clear understanding of the real significance of the third "law" of Newton, the assumption of the equality of action and reaction in the force interactions of bodies. Many of the basic ideas in Newtonian mechanics were not really clarified until the latter part of the nineteenth century, e.g., the significance of momentum and energy and the conservation principles involving them. It is not at all surprising that Newton's nonscientific contemporaries were mainly impressed by the uncanny and almost magical way in which he was able to calculate the motions of the heavenly bodies and introduce order on a much simpler basis than his predecessors. They took from his writings that which they felt they could understand, and incorporated it into their own thinking about the world. That his great achievement was not unappreciated by at least some of the literary lights of the eighteenth century is clear from Alexander Pope's famous epitaph:

> Nature and Nature's laws lay hid in night.
> God said: Let Newton be! and all was light.

In the same vein William Cowper wrote of:

> ... Newton, child like sage!
> Sagacious reader of the works of God.

The Scottish poet Allan Ramsay composed an ode on the death of Newton and inscribed it to the Royal Society, over which Newton had presided for so many years:

> Great Newton's dead, full ripe his fame.
> Cease vulgar grief, to cloud our song.
> We thank the Author of our frame
> Who lent him to the world so long.

But can one find in the actual writings of the eighteenth century an influence of Newtonian thought on the ideas expressed? This is always a speculative procedure and one beset with many pitfalls. However, it is reasonable to assume that the kind of rationalism expressed so obviously in Pope's *Essay on Man* reflects an appreciation of the impact on thought of the scientific revolution typified by Newton and his immediate predecessors and contemporaries. The same sort of thing is evident in the eighteenth-century philosophers Berkeley and Hume. The latter's skepticism indeed seems intimately related to his understanding of science. Berkeley was not so favorably affected by the calculus of Newton and Leibniz and inveighed against "infinitesimals." Still, he had to be acquainted with the new brand of mathematical analysis even to be able to attack it. Some authorities think he knew the answers to his so-called philosophical criticism and was merely having a good time in a fashion not unheard of in contemporary philosophical attacks on scientific theories. On the empirical side, Berkeley made his famous excursion into public-health matters by extolling the virtues of tar water, said to be based on experience gained in his sojourn in Rhode Island. More closely related to modern science were his attempts at a psychological theory of vision. Of course, it may be objected that the works of Berkeley and Hume are philosophy and not literature. Most humanistic scholars, however, include them in the history of English literature.

An important point often made but deserving of more emphasis is the extent to which learned humanists of the eighteenth century were acquainted with scientific theories and investigations. This is illustrated even by those who like the poet and artist William Blake attacked science as destructive of the exercise of human imagination. They missed the boat, but, as we have previously emphasized, they made what they considered to be a careful study and an honest appraisal. This is quite different from the neglect of science by many literary figures of the late nineteenth and most of the twentieth centuries.

Generalizations on the theme of the relation between science and literature are unsafe, but it is not difficult to trace in early nineteenth-century English literature the same sort of concern with science which marked the earlier period. The names of Wordsworth, Shelley and Coleridge come to mind here. Though according to Douglas Bush, "Wordsworth's thought or feeling is altogether nonscientific, and is not concerned with evidences of design or indeed with much except his own response to the idea of unity of Being,"[12] it is difficult to believe after even a cursory examination of Wordsworth's poetry that he had no appreciation for science and scientists. This is definitely belied, for example, in the sixth book of his poem "The Prelude; Or, Growth of a Poet's Mind" from which the following quotation is relevant:

> Yet may we not entirely overlook
> The pleasure gathered from the rudiments
> Of geometric science. Though advanced
> In these inquiries, with regret I speak,
> No farther than the threshold, there I found
> Both elevation and composed delight:
> With Indian awe and wonder, ignorance pleased
> With its own struggles, did I meditate
> On the relation those abstractions bear
> To Nature's laws, and by what process led,

[12] *Science and English Poetry* (London: Oxford University Press, 1950).

> Those immaterial agents bowed their heads
> Duly to serve the mind of earth-born man.
> From star to star, from kindred sphere to sphere,
> From system on to system without end.

And then further on:

> Mighty is the charm
> Of these abstractions to a mind beset
> With images, and haunted by herself,
> And specially delighted unto me
> Was that clear synthesis built up aloft
> So gracefully; even then when it appeared
> Not more than a mere plaything, or a toy
> To sense embodied: not the thing it is
> In verity, an independent world,
> Created out of pure intelligence.

It seems clear that the man who wrote these lines had a clear conception of the hypothetical and imaginative character of science and indeed a better idea of it than many of the scientists of his time. Possibly a reason for this may be found in his close friendship with the Irish mathematician, astronomer and physicist Sir William Rowan Hamilton, who rarely strayed from Dublin but succeeded in providing a new point of view in mechanics and optics which proved of enormous value in celestial mechanics and later served as a basis for the modern development by L. de Broglie of the version of quantum mechanics known as wave mechanics. Hamilton was an original genius of great imaginative power. His first important paper, *Theory of Systems of Rays,* was presented to the Royal Irish Academy in 1827 when its author was twenty-two years of age. The extension of this to the formulation of what we now call Hamiltonian dynamics came in 1834. The latter part of Hamilton's professional career was devoted largely to pure mathematics and to his invention of quaternions, closely associated with what is now known as vector analysis.

Hamilton's interests were very broad,[13] and in particular at an early age he began to write verse. It is no exaggeration to say that in the course of his life he produced reams of it! It was in connection with his ambitions in this direction that he made the acquaintance of Wordsworth and submitted his efforts to the veteran poet for criti-

[13] See, for example, the monumental biography by R. P. Graves, *Life of Sir William Rowan Hamilton* 3 vols. (Dublin University Press, 1882-1887).

cism. They became fast friends, even though the scrupulously honest Wordsworth had to advise the young Irishman that it would probably be better for him to follow science than poetry as a profession. For a time Hamilton had wavered in his decision with respect to his career. In a very real sense we may look upon Wordsworth's advice and Hamilton's somewhat reluctant agreement as a genuine service paid by the humanities to science! At the same time Hamilton's interest in writing verse continued to the end of his life and he had many discussions with Wordsworth about the nature of poetry, both by correspondence and by personal intercourse at their respective homes. Wordsworth's mastery of both the use of ordinary words and scientific terminology is manifest in his criticism of one poem of Hamilton in which the latter makes a "touch of sympathy" *whisper* and makes "Science's ray" *shine*. Wordsworth pointed out that "sympathy might whisper, but a *touch* of sympathy could not." At the same time to say that a ray shines is very awkward; we can say that the sun shines, but hardly that a ray shines since the word ray is our way of describing scientifically the propagation of light. A person who could make this kind of criticism certainly understood science!

Wordsworth's feeling about Hamilton may be summed up by the judgment he expressed to more than one person that the two most wonderful men he had ever met were Coleridge and Hamilton. This kind of association of the humanities and science was not rare in the early nineteenth century. Would we had more of it in the twentieth!

Any attempt to appraise the relation of science to literature during the later nineteenth century is complicated by two factors. The first was the rapidly growing industrial development based in part at least on the advance of science, and the second was the increasing technicality of basic science accompanied with a more elaborate terminology which posed formidable problems of understanding what the new sciences of electromagnetism and thermodynamics were all about. The first was discouraging to men of letters who were concerned about the effect of science on society and concluded that the technological influence through factory labor and the like was all to the bad, both in its social implications and in its emphasis on the material aspects of life. One gets the impression that the literary people of the middle and late nineteenth century hated things like railways and electric power partly because they disfigured the pleasant countryside and partly because of the unpleasant social problems which seemed to emerge in corrupting the good old ways. The new romantics like William Morris, Dante Gabriel Rossetti and Algernon Charles Swinburne represent this attitude very emphatically. On the other hand, Alfred Tennyson in his poetry took a more optimistic point of view, even with respect to the technological advances, and at the same time evidently appreciated at higher value the imaginative aspects of science. This is evident in his poems "Locksley Hall" and "In Memoriam." Tennyson was evidently influenced by the rise of the evolutionary concept in biology and its implications for the position of man in the universe.

On the whole, it is fair to say that the gulf between the scientist and the man of letters widened toward the end of the nineteenth century. There were, however, exceptions. H. G. Wells was trained as a scientist and narrowly escaped becoming a physicist—he was cured by a distaste for formal laboratory work in this subject and a lack of willingness to undergo the rather stiff apprenticeship needed for success in science. He wanted to know all about atoms and light without laying the groundwork of mechanics and mathematics. However, he imbibed enough of the scientific spirit to write some high-quality science fiction in the early part of the twentieth century. And the same spirit undoubtedly moved him to his later sociological novels with their emphasis on the possible perfectibility of society through technical advances and a more rational approach by man to his social problems. But Wells was exceptional in this respect and of course his position in the literary firmament was never rated very high by the critics.

One might see in the rise of the realistic school of novelists in the twentieth century an indication of the influence of science. But there is no evidence that it was based on any very wide knowledge of modern science on the part of the authors. They were concerned more and more with the seamy side of life and may have felt that this is somehow connected with the increasingly important role played by science in the affairs of men; but their mood has been defiant and pessimistic rather than optimistic about man's future. A very recent exception is Sir Charles Snow, who began his career as a physical chemist and did creditable scientific work before he turned to novel writing. Many of his novels introduce scientists as characters and tend to "humanize" the whole profession.

The subject of the presumptive influence of science on literature in the contemporary period is an enormous one, and we could not hope even to summarize it here. However, we ought to call attention to the "stream of consciousness" idea on which the philosopher and psychologist William James laid so much emphasis and which many modern novelists have adopted as a technique. James, it will be remembered, called attention to the fact that the thinking processes of ordinary human beings consist as a rule of a chaotic but continuous succession of more or less disorderly and unformed images often having little connection with each other, though suggested by curious associations which seem to arise with little or no conscious effort. It is true, of course, that steady, sustained thinking of a logical character is possible to the trained mind if the possessor of it is willing to make the effort, but this seems to be the exception as far as what may be called normal human beings are concerned. James made the point that psychologists have an obligation to study this "stream of consciousness" if they are to do justice to the problem of human thinking. Certain modern novelists have decided to take advantage of the suggestion. In particular, James Joyce in his novels employed the technique by describing at great length and in minute detail the random thoughts passing through the heads of his characters, thus providing a style of novel writing replacing the narration of action and incident and leading to the achievement of a greater sense of immediacy and a graph of what the character being described "really is thinking."

Our account so far has been limited almost entirely to English literature. If we were to aim at anything more complete we should have to concern ourselves with the similar problems in other countries of western Europe. We should have to explore, for example, the position of Voltaire in eighteenth-century France and the influence on his writings of his association with the great French scientists like D'Alembert in the preparation of the famous French Encyclopedia. We should find it necessary to devote a whole chapter to Goethe, the great German humanist, who thought deeply about science with his theory of color and his interest in evolutionary theories, and whose poems and dramatic works are full of scientific and philosophical allusions.

All this is the merest preamble to a definitive study of the influence of science on literature. But we have probably included enough to convince the skeptical that the influence has genuinely existed and that there is no reason to doubt it will continue to do so in varying measure depending on the spirit of the times.

The interesting question now arises: To what extent is the relation between science and literature a reciprocal one? In other words, what sort of influence has literature exerted on science? The story here mirrors that in the other humanities. It is hard for a physical scientist to doubt that the influence is mutual. He thinks at once of the famous principle of action and reaction of Newton, in accordance with which if particle A acts on particle B, then particle B acts back on (or reacts on) A with the same influence in the opposite direction. And so it seems to be with all human beings and all human institutions, if we probe deeply enough.

To be specific, the scientist is a creature of his cultural environment and cannot fail to show in his work traces of the impact of every part of it. His education like that of all involves acquaintance with literary creations. No matter what attitude he takes toward them they have an influence which he cannot avoid, even though he may never acknowledge his debt to it. In particular, the free use of the imagination by the poet cannot fail to encourage the speculative urge of the scientist in framing imaginative theories which in the last analysis constitute the most significant part of science. Even the hostile criticism of science by humanists, who stress the narrow specialization of scientists, is a spur to see that scientific writing becomes more intelligible. The scientist is by no means impervious to proddings by his literary colleagues and more articulate scientific writing often results from the caustic criticism that humanists have leveled at the inarticulateness of scientists. Undoubtedly, Diderot and Voltaire exercised a great influence on the writings of D'Alembert, just as Wordsworth must have done for Hamilton.

We meet here a matter of considerable importance for the relation between science and the humanities in general. It is clear that one of the causes of the misunderstanding which too often has embittered their mutual feeling has been the inability of the scientist to make clear to the humanist precisely what he is doing, in graceful and appealing language. Granted that the professional needs to invent special terminology in order to pursue his business with efficiency and economy, it remains true that he can, if he wishes, describe to the nonscientist the gist of what he is doing in the language of ordinary speech. Great scientists have often been willing to do this. The names of Huxley, Tyndall and Faraday come to mind, and the reader can supply some of more recent date. To be sure, the task is by no means an easy one and misleading impressions can be created in the so-called popularization of science by even the most conscientious of scientists. The effort, however, must be made and carried on continuously, if only to destroy the illusion which seems to persist in the minds of some humanists that in going beyond the description of that nature which is observable through the senses, the scientist is somehow distorting the perspective of the race, making man out of tune with "nature" and, in general, by introducing large realms of human activity which transcend simple sensory experience, helping to remove esthetic qualities from man's reaction to his environment.[14] It seems clear that such a viewpoint would suggest that if science *is* to have an influence on literature and the other humanities it will be deleterious. But surely this attitude is based on an unfortunate misunderstanding of the whole nature of science, which from very early times has been marked by man's insistence on his right to create new experience in addition to that presented to his unaided senses. This of course is no new thing: scientists have been doing experiments for centuries and all the experience thus created is just as much a part of nature as the observations of naturalists. From this there seems to be no escape except in sheer obscurantism. The creation of experience by the scientific method will go on as long as man persists and the spirit of curiosity lingers. What is created must be a part of life and hence subject to the investigative reaction of those who call themselves humanists. In the past they have always risen to the occasion. Who can doubt they will continue to do so?

[14] Such a point of view has been very forcefully and effectively expressed in the writings of Joseph Wood Krutch. Cf., for example, his *Human Nature and the Human Condition* (New York: Random House, 1959), and his column "If You Don't Mind My Saying So" in *The American Scholar*.

CHAPTER 4

SCIENCE AND

PHILOSOPHY

The Nature of Philosophy

Since philosophy ranks in general as one of the humanities, any discussion of its relation with science presumably belongs to the previous chapter. Its overwhelming ideological importance in human culture, however, justifies a separate chapter. This decision is reinforced by the admittedly great concern that philosophers have long expressed over the problems of science, even when it is clear they have not wholly understood the latter as scientists think these ought to be understood. On the other hand, most great scientists have sooner or later recognized the existence of philosophical questions of great significance in their scientific work. It must in all candor be confessed that when they have tried to come to grips with these matters they have not always covered themselves with glory, and have often provided shining targets for the sharp criticism of philosophers. Clearly our discussion will have to proceed with some care!

Caution is called for in more than one respect, since, as is the case with most forms of human intellectual activity, there is considerable uncertainty as to just what philosophy is. Evidently, before we take up in detail its relations with science, we need to make clear what we shall understand by its nature. Everyone knows that the name itself comes from the Greek, meaning "love of wisdom," and this presumably implies that philosophers are so enamored of wisdom that they do all in their power to achieve it. But it is not altogether clear what this wisdom is which they seek. In everyday usage it refers to the reflective ability to judge properly concerning great issues in the conduct of life, and certainly many philosophers have concerned themselves with problems of this kind. But most would doubtless wish to extend the meaning to encompass concern with the greatest questions which human beings can ask about their existence, such as: "Who am I? Where did I come from? Where am I going? What can I know? How can I reason? What ought I to do? What is the meaning of the experience I call my life?"

Now philosophers have at various times endeavored to provide answers to these questions and in so doing have constructed systems of thought of sufficient elaborateness to receive technical names such as epistemology (theory of knowledge), ontology (theory of being), ethics (theory of behavior), logic (theory of reasoning), etc. But if we seek a common ground of activity in all this we encounter difficulty. It is not surprising then that men who have called themselves philosophers have often given the name to quite different forms of intellectual activity. C. J. Ducasse[1] examines several interpretations of philosophy given in fairly recent times and criticizes them in the light of his own ideas. He dismisses Bertrand Russell's definition of philosophy as the "science of the possible" as reducing philosophy essentially to logic, a too narrow point of view. In a similar vein, he discards the view of Rudolf Carnap, the Viennese positivist, that philosophy should be nothing but the logical syntax of the language of science. Other philosophers have felt that the subject is really concerned with the systematic study of meanings (S. K. Langer) or with the practical handling of the problems of men (the instrumentalism of John Dewey). Ducasse finds fault with them all and finally winds up with his own view, which is that the essential task of philosophy is to examine appraisals or value judgments on all sorts of problems. He also feels that these appraisals are to be handled essentially as are the facts of science and that the method of philosophy is really the method of science as we have set it forth in Chapter 2, except that the basic subject matter consists not in facts of experience but in appraisals, that is, the valuations made of experience by individuals. Thus on this view philosophy tries to draw conclusions from fundamental assumptions such as "to steal is wrong" or "one must always tell the truth," etc.

[1] *Philosophy as a Science* (New York: Oskar Piest, 1941).

Ducasse's view is particularly interesting from the standpoint of the relation between science and philosophy. But we must be careful again to realize that the definition of an activity like philosophy is a tricky business. There is a natural tendency for a thoughtful philosopher to define it as he thinks it *ought* to be. This may have little relation to what other persons called philosophers have actually been doing. Somewhat more reasonable would seem to be what we did in the case of science. There, in effect, we examined what certain people called scientists actually have done and are doing, and tried to find the essential elements common to the work of all. This is difficult enough in science, but it is in all likelihood far more difficult in the case of philosophy.

There is indeed one aspect of most philosophical works which has always attracted the attention of scientists, and physical scientists in particular, because it has always seemed to imply a weakness in philosophizing and perhaps to explain in part why philosophers in spite of their great mental effort appear to have accomplished so little. While science abstracts from the totality of experience limited domains for intensive study, philosophers wish to take the whole of experience, or, in short, the universe for their field of investigation. Scientists have learned (so they think, at any rate) that it is difficult to make useful generalizations about the universe, but philosophers are not content with less. This reminds one of the story that is often told (among physical scientists) about the relation of mathematics and physics respectively to philosophy. It goes this way: Mathematicians have for ages past given much attention to the mysteries of Nature and by dint of great and unremitting effort have at length learned almost everything about nothing; at the same time the philosophers have not been idle in their endeavor to penetrate the puzzle of existence and as a result have finally learned almost nothing about everything. Finally, the physicists, humble folk, have not set their aspirations quite so high and thus have succeeded ultimately in learning a little something about something. The moral of this little allegory should be obvious.

And yet there must be some reason why men have not been contented to forego the larger questions; all existence seems such a mystery that it literally cries out for interpretation and the direct description of science seems so limited. Moreover, science leaves untouched so many vital matters; for example, how can we be sure that all our scientific knowledge is not a mere illusion? Is there a real world out there waiting to be observed and described, or is it all in our imagination? Someone is bound to ask such questions and to seek answers out of his own imaginative contemplation. These endeavors are to be respected as part of the endeavor of man to understand the meaning of his existence.

Enough has been said to justify the view that it is impossible to provide a pinpoint definition of philosophy. There are many philosophies just as there have been many philosophers. If there is a coherent thread running through them all it is the attempt to come to grips fundamentally with all aspects of man's attitudes toward his experience and to do this with due regard to what is commonly called logical reasoning. More specifically, when a philosopher decides to adopt a certain set of attitudes toward human experience as a basis for his theory of what *really* goes on in the world, he then desires to draw in the most logical manner possible all the conclusions possible and to show that they can be identified with what people have actually observed, thought or felt about experience.

Thus the philosopher wedded to realism as a doctrine makes the assumption that man exists in a world of objects which actually exist independently of the presence of man. He then proceeds to draw all the conclusions he can from this fundamental assumption and shows to his satisfaction that these agree with all that man experiences and says about his experiences. On the other hand, the philosophical idealist insists that it is a much better assumption that man's whole experience is nothing but a body of ideas in his mind; there are no such things as physical objects, there are only ideas which we name such. Once again he proceeds to draw consequences and to satisfy himself that these provide a complete identification with experience.

It is all a very fascinating game full of subtle hazards and surprising pitfalls, but leading for many clever individuals to what they feel are illuminating revelations of reality. Our problem is to see what relation all this has to science. Has philosophy had an influence on the development of science, and what role does it play in present-day science? Do scientific developments influence philosophical views and what impact does this make on society as a whole? We now give attention to these questions.

Philosophy and the Logical Structure of Scientific Theories

In facing the problem of the role of philosophy in science we are irresistibly led back to the logical nature of scientific theories as developed in Chapter 2. In numerous places there we were led to comment "this is a philosophical question." For example, in the discussion of the success of a theory we mentioned the earlier desire to talk about the "truth" of a theory as something to be established by testing its predictions experimentally. But we had to admit that the concept of truth is a difficult one. Its meaning has been argued for ages by philosophers without ultimate agreement. So right here we have a philosophical problem arising in the heart of the examination of the meaning of what a scientist does. It will be recalled that we decided to side-step the problem by discarding the concept of "truth" in science and replacing it by the notion of "success." We felt we could render the latter sufficiently specific to make it useful. But the philosopher might object that we are merely trying to solve a difficult problem by changing the terminology. If the scientist retorts that he has no time to be more than pragmatic in such matters, the philosopher might still gently chide him for supposing that his notion of "success" is the best possible way to justify a scientific theory. Perhaps a more thorough study of "truth" might establish scientific theories more effectively. We have no intention of taking sides in such a controversy. It is sufficient for our purpose to point out the perfectly valid possibility of the existence of such a difference of opinion, and in particular to admit that the scientist's choice of "success" in preference to "truth" is a kind of value judgment which is, of course, philosophical in character. Whether the scientist likes it or not, in the course of theorizing he has to make philosophical appraisals and value judgments. It is only natural to ask if it would not be much to the point for him to make a careful study of what professional philosophers have to say about fundamental concepts, and if he could not have much to gain from discussions with philosophers. The answer appears to be in the affirmative. In fact, most great scientists sooner or later tackle vigorously the philosophical aspects of their work. We need mention here only men like Bohr, Einstein, Planck and Eddingtion, all of whom will figure largely in specific philosophical questions to be discussed in detail presently.

Other illustrations of philosophical problems in science occur in

Chapter 2. Consider the problem of the nature of scientific concepts and hypotheses. Should these be based on the fundamental assumption that there exists a real external world, which is close in its character to the world of our experience? If so, it would seem only sensible to construct concepts and frame hypotheses which are suggested as directly as possible by the objects and processes we observe. On the other hand, if the scientist adopts the idealistic viewpoint and does not believe in the objective existence of a real world external to experience, presumably he will feel himself permitted to call on his imagination freely in the building of theories. In either case, he is functioning to a certain extent as a philosopher and might legitimately ask for advice from the professionals, who could conceivably keep him from going astray in certain directions. For example, the professional might well ask just what is this "imagination" which the scientist calls to his aid and where does it come from? Is it simply a function of individual experience, or is there something mysteriously innate about it? The scientist is apt to dismiss such questions impatiently; some folks get clever ideas and that is all there is to it. The philosopher is entitled to rejoin: "Do you honestly think so?" The scientist may still be unconvinced and reply that if anyone is ever to understand where the imaginative scientific ideas have come from in the past and are likely to come from in the future, it will be the psychologist and not the philosopher. At this the philosopher probably merely smiles.

The reader will recall that in our logical analysis of scientific theories the first category consisted of primitive, undefined notions. These always trouble the philosopher, who dislikes to see elaborate structures reared on shaky foundations. Hence, while a physicist is apt to look upon such primitives as space and time from a pragmatic point of view, the philosopher wants to probe more deeply. Now he may not actually get very far in the sense of turning up something specifically useful for science. But it must not necessarily be concluded that therefore his probing is valueless; it may stimulate the scientist to look more carefully at the foundations of his theories and this can lead to results of the utmost importance. This was true, for example, when Einstein decided to re-examine the use of the concept of time in physics and was led to the fundamental transformation equations of special relativity. It may be objected that these do not really touch the philosophical question "What *is* time?" but rather somewhat arbitrarily introduce a symbol t into the hypotheses of mechanics as a mere mathematical parameter of which quantities like displacement, velocity and acceleration can be functions. This is what a physicist means by time on the constitutive level. On the epistemic level, he introduces a piece of apparatus called a clock and identifies t with the various successive coincidences of the hands of the clock with the scale on the face. What Einstein and the other developers of the theory of relativity did was to show that in order to be consistent with the observed facts of physics, if one uses the symbol t to describe the motion of a particle in one system of coordinates, then one may *not* consistently use the same symbol to describe its motion with respect to a system of coordinates moving with respect to the first with constant speed v. In fact, one must use t′, which is given in terms of t by the equation

$$t' = \frac{t - vx/c^2}{\sqrt{1 - v^2/c^2}}$$

where c is the velocity of light in free space, i.e., 3×10^{10} cm/sec, one of the fundamental constants of nature. From the equation, coupled with the corresponding transformation equations for the coordinates (x,y,z) of the particle in space, one can deduce interesting consequences, e.g., a moving clock runs slow when observed in the system with respect to which it is moving. Such results have stirred up philosophers immensely and much has appeared in the philosophical literature about the "paradoxes" of relativity. These paradoxes probably have their origin in the tendency of philosophers to interpret time in some absolute sense not contemplated by scientists. It may well be that an interpretation of time as absolute duration, something which can serve to denote change in all experience including that in living organisms, is not compatible with the more restricted view of time used by physical scientists. This, however, is a problem of considerable importance and difficulty and will doubtless continue to be studied by both scientists and philosophers.[2] Our purpose here

[2] See, for example, the discussion of the significance of time in physics in Lindsay and Margenau, *Foundations of Physics* (New York: Dover Publications, 1957), pp. 72 ff. More recent works are G. J. Whitrow, *The Natural Philosophy of Time* (London: Nelson, 1961), and R. Schlegel, *Time and the Physical World* (Michigan State University Press, 1961).

has been to give another illustration of the philosophical questions that confront one at every turn in the logical analysis of a scientific theory.

It will pay to take up another example that has caused much discussion among scientists and philosophers. In our analysis of scientific theory we stressed the rather arbitrary character of the hypotheses which scientists invent out of their imaginative power, and the somewhat marvelous result that these things actually do serve to describe and even predict experience. What we did not stress was the fact that in developing concepts and hypotheses we utilize more or less unconsciously methods of thinking which human beings have somehow evolved through the ages and which are now assumed to be the proper methods for gaining knowledge. For example, we use the principles of Aristotelian logic as the distillation of common sense reasoning, and we have no hesitation in applying the basic postulates of Euclid about magnitude, congruence and order. In a certain sense therefore our description of experience is conditioned by our mental habits. Some might go so far as to say that our whole science is epistemological in character, i.e., is tailored to fit the processes of our minds. Scientists with philosophical leanings toward idealism are naturally sympathetic with such views. One of the outstanding examples is Arthur Stanley Eddington, who besides being an outstanding astrophysicist was a leading authority on the philosophy and methodology of physics. After long meditation on physical theorizing, and undoubtedly carried away by the successes of mathematical physics, he finally reached the conclusion that the whole universe of experience is really the creation of man's mind. Following is the famous statement which makes clear his standpoint:

> Unless the structure of the nucleus has a surprise in store for us, the conclusion seems plain—there is nothing in the whole system of laws of physics that cannot be deduced unambiguously from epistemological considerations. An intelligence unacquainted with our universe, but acquainted with the system of thought by which the human mind interprets to itself the content of its sensory experience, should be able to attain all the knowledge of physics that we have attained by experiment. He would not deduce the particular events and objects of our experience, but he would deduce the generalizations we have based upon them. For example, he would infer the existence and properties of radium, but not the dimensions of the earth.

The mind which tried simultaneously to apprehend the complexity of the universe would be overwhelmed. Experience must be dealt with in bits; then a system must be devised for reconnecting the bits, and so on . . . . In the end what we comprehend about the universe is precisely that which we put into the universe to make it comprehensible.[3]

This (doubtless colored by the views of Descartes, Berkeley and Kant) is probably the extreme version of what may be called the epistemological theory of the methodology of science. It has of course given rise to an immense amount of controversy. To empiricists, i.e., those who stress the experiential basis of all scientific knowledge, it makes little sense and they have attacked it vigorously. The most reasonable attitude to take toward Eddington's view is that it possesses a grain of plausibility but must not be swallowed whole. It is certainly the case that some scientific theories created by human imagination have been remarkably successful, but we also know hosts of others that are in the dustbin because they wholly failed to describe experience successfully. How can we *a priori* be sure that any theory will uniquely describe experience? The answer is we cannot, except by testing it against actual experience. On this basis Eddington's plan becomes a pious gesture whose originally rather exciting possibilities fade away in the light of the actual situation in science.

Eddington's idea does suggest, however, a brief examination of some other philosophical ideas which have influenced scientific thinking in some respects. One of these is the notion of *purpose* or the teleological concept. We mentioned this briefly in the discussion of value judgments in Chapter 3, where we contrasted the Newtonian view of mechanics with that of Hamilton. In the former, the behavior of a dynamical system is obtained from the solution of a set of differential equations of the second order and the state of the system at any instant is found by substituting appropriate initial conditions into the solution. In the latter, the time integral of a certain function of the energy of the system is set up, and it is then required that this should have a stationary value (usually a minimum) with respect to all possible motions of the system between an initial and a final state. The carrying out of the mathematical condition yields the differential equations of motion. It is commonly said (as it certainly was believed by Maupertuis) that whoever designed the universe purposed that this particular integral should have a stationary value. From the standpoint of the formal mathematical development of mechanics, it is of course not necessary to assign this philosophical interpretation. The Hamiltonian scheme is simply an alternative way of setting up the basic hypotheses of mechanics. But some people insist on assigning larger meaning to what others consider merely formal hypotheses.

This example from physics may not be considered by philosophers to be a very good one. Biology provides a more appropriate milieu for the notion of purpose. The growth of a complete and complicated organism from a single cell poses theoretical problems of great difficulty, and it is not surprising that many have sought to bypass these by seeing built into the germ cell the purpose of becoming or emerging into the appropriate adult organism. On this view the cell has a goal in view and behaves in its development in such a fashion as to achieve its end. This in essence is the standpoint of the modern preformationist, who believes that the germ cell contains all the information necessary to guide it to the adult form. In fact, the earlier preformationists held to the view that the complete organism was present in miniature in the cell and needed only nourishment for growth to the adult size. In opposition to this view is epigenesis which holds that the embryo starts in a very simple form and then differentiates as it grows. It creates, as it were, in the process of growth the information needed to guide its future development. This point of view is of course the one most directly substantiated by experiment, as no way has been created to locate in the fertilized ovum the complete pattern or blueprint of its future course. Still, both viewpoints leave unsolved the problem of how it happens that the ovum of a cat for example knows enough to develop into a cat. The need that some feel for a notion like purpose or goal-striving appears pretty obvious. We do not intend to argue it here, but merely to call attention to it as another example of the incursion of philosophical ideas into science.

As a final illustration in the present context we may mention the concept of *causality*. The notion that nothing happens in human experience unless something else has happened previously in time so closely connected with the event in question that unless it had taken place the given event would not have taken place, is rooted in human thinking on even very primitive levels. No wonder then that philosophers have felt it their duty to introduce the notions of cause and effect into all phenomena. It has seemed natural to them to interpret from this point of view scientific laws which describe regularities in experience. Thus Boyle's law, which we have already used as a stock illustration in Chapter 2, can be looked upon as a case of cause and effect in the sense that when one increases the external pressure on the gas (the cause) the effect produced is a decrease in volume (if the temperature remains constant). This sounds sensible until we reflect that we might just as logically have looked upon the decrease in volume as the cause and the increase in pressure as the effect. All that the law does is to relate two physical quantities susceptible of both constitutive and epistemic definition. No idea of enforcement of behavior or necessity is involved and to insert one provides no assistance whatever in the scientific study of the behavior of gases, though it may conceivably make the philosopher or nonscientist happier.

Does this mean that no notion of causality, related somehow to the naïve conception of cause and effect, can play a useful role in science? By no means! In scientific theory the idea of causality enters with the very idea of the existence of scientific law as a description of a pattern or routine of experience. Causality in science then appears as an expression of the uniformity we find in nature or somehow manage to impose by our method of observation and thinking on our experience, a pattern as opposed to confusion and chaos. By this kind of modest restriction of the concept, the scientist is spared the necessity of worrying about "the first cause," so dear to the hearts of metaphysicians.

Now it must be confessed that there is some disagreement among scientists about the concept of causality. Among many it is essentially equivalent to the notion of *determinism*. We have pointed out that in mechanics, for example, once we have given the initial state of a dynamical system, i.e., know the position and velocity of every particle of the system, and know the forces which act on all the particles, we can unambiguously predict the state of the system at any future instant of time (or any past one, for that matter). In other words, once the state of the system is known at some chosen instant, the states at all other times are precisely determined. A standard popular example is the ability of the astronomer to predict the precise time at which the sun, earth and moon will be so arranged with respect to each other as to lead to a solar eclipse. It might be thought that this is of necessity tied to the existence of regular laws and hence causality (as we have interpreted it) and determinism amount to the same thing. However, the situation is not quite so simple. In the kinetic theory of gases, for example, the number of molecules supposed to constitute a gas is so enormous that, although formally and in principle one should be able to solve the equations of motion for all particles and by imposing appropriate initial conditions determine precisely what every single molecule is doing at any assigned time, in fact we are unable to carry out the program as it is too complicated and runs into mathematical difficulties which have never yet been overcome. What we do is to fall back on a method called statistical (hinted at in Chapter 2), whereby we limit our calculations to average quantities based on the probable behavior of the particles. Thus the average effect of the bombardment of the very large number of molecules on the walls of the container of the gas will be looked upon as corresponding to the observed pressure of the gas, etc. This type of theory gives up the attempt to achieve determinism with respect to all constituents of the system but does succeed in setting up relations among average quantities which are of the same validity as other scientific laws and can be tested experimentally to see whether they stand up as descriptions of routines of experience. They therefore preserve the causal concept as we have defined it. But on the finer level of the behavior of the individual molecules they are indeterministic. So we may say that statistical laws are causal in character but indeterministic.

We meet this situation in more drastic form in modern quantum mechanics, where the future behavior of individual atoms is considered in principle (i.e., by direct postulate) undetermined. (See Chapter 3.) Nevertheless, the theory of quantum mechanics leads to equations implying regularities in atomic systems which can be tested experimentally. On this basis quantum mechanics preserves causality, though it abandons determinism. We shall return to this below since it has been a source of confusion in popular philosophical presentations of quantum theory.

Though we have of course done little more than scratch the surface, enough has been said to give some conception of the extent to which notions commonly called philosophical and which philosophers have been discussing from time immemorial have played a role in the development of science. It is now time to turn to a consideration of ways in which science has influenced philosophy. This will provide a fairer picture of their mutual relations.

Science and Logical Positivism

In seeking for an example of the reciprocal influence of science on philosophy we might of course go back to ancient times when there was no real difference between the two disciplines. Many of the early philosophers were also scientists and certainly their attempts to understand the nature of experience and describe the world in which they lived colored the philosophical views they developed. Without question Aristotle's metaphysics took much of its form from the ideas of nature expounded in the *Physics* which preceded it, though whether the mere order of occurrence of the books (the *Metaphysics* came directly after the *Physics* in the earliest compilation of his writings) has any substantial significance may be doubted.

However, we do not wish here to get tangled up with questions of the ancient history of science and philosophy. It will suffice for our purposes if we devote some attention to a more modern influence of science on philosophy, i.e., the rise of logical positivism, more recently termed logical empiricism. In the city of Vienna in Austria there grew up in the second decade of the twentieth century a group of philosophers and philosophically minded scientists (mainly physicists, logicians and mathematicians) who were much impressed on the one hand with what they considered to be the bankruptcy of professional philosophy as represented by the classical systems of metaphysics, and on the other hand with the successes of physical science as represented by mathematical theories like relativity and the quantum theory. They were greatly influenced also by the viewpoints of men like Ernst Mach, who felt strongly the empirical character of science, and, though not disbelievers in theory, emphasized the need to test all theoretical predictions by thoroughgoing experimentation. Mach indeed may be considered the initiator of the group. This group, at first referred to as the Vienna circle (Wienerkreis) became known ultimately as logical positivists through their insistence that no philosophy was worth anything unless it contained within it assertions susceptible of actual verification in experience. Thus, they took over a cardinal principle of the method of science and insisted it should also apply to philosophical reasoning. The Vienna circle came to include physicists like Moritz Schlick and Philipp Frank, the mathematician H. Hahn, the philosopher Rudolf Carnap and the political economist Otto Neurath. It had considerable influence on the intellectual development of theoretical physicists like E. Schrödinger and mathematicians like R. von Mises and K. Gödel. The unhappy events of the thirties in central Europe followed by World War II broke up the circle. Many of the members are now in the United States and are still active, notably Frank and Carnap.[4]

The fundamental idea of logical positivism obviously has a devastating effect on metaphysics, which in the classical tradition has dealt freely with propositions not susceptible to experimental verification. Thus, according to Immanuel Kant, the axioms of Euclidean geometry are *a priori* synthetic propositions, i.e., concepts imposed on the mind of man of such a nature that he cannot think at all logically about space relations without employing them. But Poincaré and others showed convincingly that there is no experimental way of verifying such a proposition. Then there were the realistic philosophers who built metaphysical systems on the basis of the assumption of the existence of a real world of physical objects. They have been opposed by the idealists who maintained vigorously that all experience is created by the mind. To the logical positivist all such speculation is meaningless and useless: there is no possible way to test it against the facts of experience themselves. Therefore, why should people concern themselves with it?

Thus, the logical positivist, who might more logically be termed (as indeed he has been) a logical empiricist, pretty thoroughly disposes of metaphysics as well as large sections of classical philosophy in the fields of epistemology and ethics. Indeed, Rudolf Carnap ul-

---

[4] A good review of logical positivism and the Vienna circle may be found in Philipp Frank, *Between Physics and Philosophy* (Harvard University Press, 1941).

timately came to the view that the only use left to philosophy was to serve as the logical analysis of the language of science.[5] This would essentially reduce philosophy to logic. Logical analysis means the study of the forms of expression used in scientific description. It is equivalent to all intents and purposes to what is usually called syntax by the students of language. Now language is certainly one of the greatest tools ever devised by man, since man is essentially a communicating animal. The study of it is therefore of the highest importance in any estimate of the role of science. We shall go into this in some detail in Chapter 6. But to reduce philosophers to linguists scarcely flattered persons who have felt that they have been meditating on some of the most vital questions relating to man's life and destiny. Most professional philosophers have not taken kindly to logical positivism, but there is no question about its influence on the development of modern philosophy. Even in the process of criticizing Carnap's version of logical positivism, as Ducasse, for example, has done in incisive fashion,[6] philosophers have been forced to think more clearly about what philosophy ought to be expected to accomplish as a humanistic discipline, and to attempt programs better calculated to make closer contact with the world of experience. This is well illustrated, as a matter of fact, by the program outlined by Ducasse, as has already been mentioned.

The Role of Philosophy in the Interpretation of Science. The Cultural Lag

One aspect of philosophy we have not yet emphasized sufficiently, although we have indeed hinted at it in Chapter 1, is its part in interpreting the concepts of science to the general public. The chief results of science are usually expressed in highly technical language, which most nonscientists are unwilling to learn. On the other hand, the popularization of science by scientists, with a few notable exceptions, has not been conspicuously successful. In many cases it has actually served to increase the misconceptions about science which are so common among the cultured public, tending as it has in general to stress the factual and technological aspects of science, which

[6] Rudolf Carnap, *The Logical Syntax of Language* (London: Kegan Paul, Trench and Trubner, 1937).
[6] Cf. C. J. Ducasse, *op. cit.*, Chapter 7.

people think they can get hold of more readily. Yet, as we have stressed in Chapter 1, probably the greatest influence of science on society has come through its ideas. Now the people who study and play with ideas more than anyone else are the philosophers. The expectation therefore follows that philosophers would study the ideas and methods of scientists, and by saying what they think their ultimate significance is for society, would serve ultimately as interpreters of science in the language of ordinary speech to the lay public.

This expectation is indeed borne out by an examination of the writing of many of the great philosophers. Kant, for example, in his *Critique of Pure Reason* presented the principle of causality or the so-called law of cause and effect in such a way as to bring out what he thought scientists meant by this idea. He thereby through his readers and philosophical disciples (who thought they understood him) disseminated among cultured people the scientific notion of cause and effect more effectively than could possibly have been done by the direct reading of the works of Newton and his distinguished scientific successors, e.g., Lagrange, Laplace, Euler and others. Moreover, the problem in any case was not so much one of trying to translate into simple and understandable terms the abstract symbolism of mathematical physics. It was rather the task of making clear the implications for thinking people in general of the great success of the scientific method as practiced by these men; of producing an increased respect for the mind of man which could somehow see an orderly pattern in physical experience instead of chaos and confusion, and do it without recourse to superstitious fancies and irrational dogmas. In a larger sense, it was the problem of getting men used to taking what we nowadays call a more enlightened and rational view of the universe.

Kant also felt it his duty as a philosopher to interpret and clarify the notions of space and time as used by the great scientists of his time and the immediately preceding period. He of course realized the overwhelming importance of these categories in the development of physical theories and assumed the obligation of analyzing them as carefully as possible. He ultimately reached the conclusion that the notions of space and time are effectively modes of thought forced on the mind and without which it could not cope with experience at all. As we have just seen this led Kant to the view that the postulates of Euclidian geometry are *a priori* synthetic judgments—we do not merely assume them; they are forced on our minds so emphatically that we cannot do geometry without them. This view seemed reasonable in the late eighteenth century and certainly carried great weight as a philosophical interpretation of mathematics and physics. It went down, of course, with the invention of non-Euclidian geometry by Wolfgang Bolyai and N. I. Lobachevski, and their demonstration that this geometry is just as logically consistent as Euclid's. Later, Poincaré showed that there is no logical way of ascertaining unambiguously whether Euclidian or non-Euclidian geometry is the "correct" one to be used in scientific investigations. In this case, the philosophical interpretation of the scientific ideas turned out to be unsuccessful in the long run. This has been an all too common occurrence, but it does not render the attempt unfruitful or its consideration unnecessary.

Another good illustration of the philosophical interpretation of scientific ideas is provided by Voltaire's *Letters on the English*. The reader of these will find in them a tolerably accurate popular account of Newton's cosmology and in particular the importance of his use of the "law" of universal gravitation, as well as his views on the dispersion of light. This description compares favorably with present-day popular versions of modern science. Voltaire was a skillful and graceful writer, to put it mildly, but more important for his contemporary readers than a superficial understanding of Newton's accomplishments was the emphasis laid by Voltaire on the success of the Newtonian method, that is, the bold and ingenious exploitation by powerful mathematical techniques of an imaginative hypothesis. Voltaire was a daring thinker himself and clearly appreciated this trait in a scientist. At the same time he was sufficiently pragmatic to recognize a successful scheme when he saw it. Thus, he preferred Newton's ideas to those of Descartes, his own countryman, since the latter's notions in cosmological physics (vortices in a continuum, etc.), while ingenious, predicted nothing that could be tested decisively.

It must be confessed that the task of the philosopher as an interpreter of science is indeed a somewhat onerous one and becomes more so as science grows and its language becomes more technical. It is not surprising therefore that there usually exists a perceptible time lag between a given scientific result deemed by scientists to have great significance and the interpretation of it in fundamental philosophical terms. Most philosophers have neither the technical equipment nor the patience to study the latest scientific discoveries in detail. By and large, they have had to depend on what scientific popularizers say about them. This has the unfortunate effect that by the time a given scientific theory or group of theories has been examined philosophically and the results of the examination have been given to the world, the prevailing atmosphere of science may have changed.

In so far as the philosophers of the eighteenth and nineteenth centuries were concerned with science, and certainly men like Hume and Kant were deeply read in these things, it was largely the physical science of Newton which influenced them and which therefore they served to interpret philosophically. In the meantime, however, science had moved on to new concepts, notably in the nineteenth century, the concept of the field, culminating in the idea of the electromagnetic field as fundamental for the transmission of light. It seems fair to say that in the nineteenth century professional philosophy pretty much lost touch with the development of science and hence was hardly in a position to provide much of an interpretation of it. This was particularly the case with German philosophy during this period, bound up as it was with romantic idealism and rather fuzzy metaphysics. An exception was indeed provided by the rise of the pragmatic philosophy in the United States, associated with the ideas of Charles Sanders Peirce, one of the most brilliant thinkers this country has ever produced. Peirce was pretty well acquainted with the scientific developments of the late nineteenth century and certainly reflected this knowledge in his writings, which however had relatively little influence at that time, though later this thread was picked up by William James and promoted vigorously. But James was really a biological scientist turned philosopher.

This lag in the philosophical appreciation of a given set of scientific ideas has become of striking and vital significance in the twentieth century. It is a well-known fact that both among philosophers and many scientists (obviously not the progressive ones!) at the end of the nineteenth century there was a rather complacent feeling that all physical phenomena could somehow be subsumed under the physical theories that had been constructed earlier in that century. One heard it said, for example, that the further task of physical science was largely that of more precise measurement, to evaluate the constants of nature to the next decimal point, etc. Of course, the persons who took this point of view turned out to be wrong, since the turn of the century saw the invention of the quantum theory and the rise of the doctrine of relativity. These, when applied to the new experimental discoveries in what we now call atomic and nuclear physics, set the stage for a new assessment of physical science. It was not indeed a change in fundamental scientific method, but rather a replacement of existing theories by new and more imaginative ones. This naturally led to considerable controversy between the more conservative scientists who felt that the old classical ideas were perfectly good and should be applicable to the new discoveries, and the more radical ones who considered that the new points of view were necessary even if it was hard to interpret them in terms of so-called "common sense." This left the philosophers in a difficult situation and it is not surprising that the philosophical interpretation of quantum theory and relativity has lagged well behind the scientific developments and in particular the technological applications.

If we except the work of the logical positivists already mentioned in a previous section and that of logicians well acquainted technically with developments in physical science like Bertrand Russell and Alfred North Whitehead, most of the professional philosophical commentary on quantum and relativity physics during the first half of the twentieth century was limited to pointing out so-called paradoxes which it was felt made these theories unintelligible from the standpoint of common sense. Thus, to the theory of relativity it was objected that any object must have a definite "real" length and that to claim, as the theory was interpreted as claiming, that its length could have different values in different coordinate systems was patently absurd and contrary to common sense. (Actually, it may be observed parenthetically, many physicists took a dim view of relativity for precisely the same reason.) The precise significance of the relativity of length eluded the understanding of the believer in a *real* universe and perhaps he was not to be blamed for this. The rather unhappy result was that the general public got the impression that not only was the theory impossibly difficult to understand by virtue of its mathematical complexity and abstractness, but in any case it had no relation to the real world, since it contradicted common sense. This could only serve to strengthen the view that contemporary physical science was something promising little or no influence on the thinking and behavior of the ordinary run of mankind. Certainly it provided no clue to the enormously significant technological developments associated with relativistic physics which were to take place before the century was half over.

In somewhat similar fashion, the interpretation of the significance of the concepts of quantum physics by professional philosophers has lagged far behind the technical development of the subject. Here again it is unfortunately for the most part misinterpretations that we have to deal with. For example, from the lay point of view perhaps the most striking conceptual consequence of the theory of quantum mechanics is the indeterminacy principle, already mentioned in Chapter 3. As we saw, this is interpreted by many if not most physicists to mean that quantum physics is intrinsically indeterministic in character. Those philosophers who confuse causality with determinism have thereby been tempted to believe that modern physics has abandoned the causal point of view. If this were so it would obviously have great philosophical consequences and some philosophers have made the most of this by exhuming the famous old problem of free will, endeavoring to find support for this doctrine in theoretical physics. Actually, the indeterminacy principle does not imply that physics has abandoned causality in the sense in which this concept is useful in physics. The really cautious thinker will think twice before he is willing to make the leap from the behavior of single atoms to the enormously complicated cooperative phenomena associated with the behavior of the human organism. But the popular mind finds it relatively easy to be confused when philosophers cite science as a support for their favorite theses.

To be fair we should not confine our castigation to the philosophers for their misinterpretations of science. Many scientists have felt the urge to philosophize and have not always thereby aided the popular understanding. We have already mentioned Eddington's ideas in an earlier section, and in particular his feeling for the epistemological character of scientific explanation. On the side of quantum physics we might cite Bohr's principle of complementarity, which was introduced as a kind of generalization of the principle of indeterminacy. It will be recalled that the latter principle can be interpreted to mean that when we try to give a spatial description of the behavior of a particle, i.e., fix its position in some reference system, we lose control of its momentum, that is, one of the variables in the dynamical description of its motion. The correlative indeterminacy principle governing energy and time says in similar fashion that when we try to give a temporal description of the behavior of the particle, we lose control of its energy. It is as if the space-time description is somehow complementary to the momentum-energy (or dynamical) description in the sense that one complements the other, but we cannot have both simultaneously with complete precision. Precise space-time location of the particle is incompatible with precise momentum-energy description.

Bohr has stressed that there is really no such thing as a single complete description of any atomic phenomenon. In his view, there exist only complementary descriptions which are mutually exclusive. To consider further the case of an atomic particle, it is well known that such particles exhibit in general wave properties and these can be detected by interference phenomena if the particles are moving fast enough. However, it turns out that if an experiment is arranged so as to fix the position of a particle at a precise instant of time the wave properties disappear, whereas, if we arrange the experiment so as to bring out the wave properties it is no longer possible to fix the position of the particle, i.e., it does not act like a particle in the sense of classical physics. From this point of view the particle and wave descriptions are complementary methods of describing the phenomena in question. It then becomes no longer a logical contradiction to employ both wave and particle concepts in the description of the behavior of an atomic entity like an electron or a proton. We need both concepts, but when we use the one in thoroughgoing fashion we are excluded from using the other.

This is certainly a rather seductive point of view, and Bohr has given many ingenious arguments in support of it. It is so tantalizing that it is not at all surprising that he was led to seek further illustrations of it outside of physical science. For example, Bohr has found it a useful device for making sense out of the puzzle of life.

One might say that there are two ways of trying to find out what the life of a living organism really means. In the first place, one can study the behavior of the organism as a whole either by observing passively its reaction to the environment or by providing appropriate large scale stimuli and noting the responses. In neither case do we gain much insight into the role which the constituent parts of the organism play in its behavior, even though we may be convinced that without them the organism could not display that large scale behavior which we call living. In the second place, we can begin to analyze the organism piece by piece, i.e., take it apart to study the function of each part. But the more detailed our analysis, the less "life" is left in the organism, so that when the study is complete, the organism is dead. We may say that the two modes of description are complementary. We need both the analytical, microscopic search as well as the macroscopic description, but they are mutually exclusive.

Bohr would go even further and apply the complementarity idea to human relations. Thus, he would look upon the concepts of justice and love in human affairs as complementary notions. The more the one prevails in a given circumstance, the less applicable is the other. The ingenuity of the idea is evident and some philosophers may find in it a unifying property. At the same time one cannot help but notice a certain elusiveness about it and a lack of pragmatic help in understanding what really goes on. It seems to promise more than it fulfills. At any rate, it provides an illustration of the way in which an idea having its genesis in a scientific theory can be given a wider interpretation which may properly be called philosophical, and hence make an impact on human thinking in general far beyond its specific scientific implications.

That such extensions of scientific concepts into broader meanings are not without their dangers will be made clear in the next section.

Dogmatism in Science and Philosophy

As a windup of this brief survey of the relations between science and philosophy let us consider the problem of dogmatism. We all know this means the rigid adherence to a certain set of principles and the positive assertion of their truth without regard for evidence against them. In the extreme case it is accompanied by a complete unwillingness to examine contrary evidence or to maintain judicious doubts pending the acquirement of further evidence.

In our discussion of the verification of scientific theories we laid stress on the unsuitability of the word "truth" in that connection, and showed a preference for the use of the word "success." However, it is not at all surprising that the author of a successful scientific theory and his disciples may become so enthusiastic that they come to believe in the ultimate "truth" of the theory in some absolute philosophic sense. There then arises a tendency to maintain the integrity of the theory in the face of all criticism and all new evidence tending to cast doubt on its appropriateness. When this happens, scientists become dogmatic and in this respect not unlike philosophers and theologians, who usually get the credit for dogmatism.

A few illustrations are in order. Though it would be injudicious to accuse Newton of dogmatism in connection with his adherence to his version of the corpuscular theory of light, he made it seem so plausible and so forcefully dismissed the claims of the rival wave hypothesis that his disciples undoubtedly felt that there could be little question of the underlying truth of the corpuscular view. The phenomenal success of the Newtonian theory of gravitation in its applications to cosmology undoubtedly reinforced their confidence. As is well known, Newton's views prevailed during the better part of the eighteenth century, though he died in 1727. Even when Thomas Young revived the wave hypothesis in the early 1800's on the basis of his famous two-slit interference experiment, it was hard for him to get a hearing, so strongly intrenched was the corpuscular theory. In the newly founded *Edinburgh Review,* a reviewer of Young's lectures attacked them in the following terms (in the endeavor to prevent the publication of Young's researches in the *Philosophical Transactions of the Royal Society*):

> Has the Royal Society degraded its publications into bulletins of new and fashionable theories for the ladies of the Royal Institution [where Young was a lecturer]? Let the Professor continue to amuse his audience with an endless variety of such harmless trifles, but in the name of science let them not find admittance into that venerable repository which contains the works of Newton and Boyle and Cavendish and Maskelyne and Herschel.

There followed more in the same vein; in particular, taunts that Young was forever changing the precise character of his hypotheses, and was after all merely ringing changes on hints thrown out in Newton's *Optics*. Thus, after an attempt at a professional criticism of Young's work which is now admitted to have little scientific justification, the author of the *Edinburgh Review* article continues:

> . . . perpetual fluctuation and change of ground is the common lot of theorists. An hypothesis which is assumed from a fanciful analogy or adopted from its apparent capacity of explaining certain appearances, must always be varied as new facts occur, and must be kept alive by a repetition of the same process of touching and retouching, of successive accommodation and adaptation, to which it originally owed its puny and contemptible existence. But the making of an hypothesis is not the discovery of a truth. It is a mere sporting with the subject; it is a sham fight which may amuse in the moment of idleness and relaxation, but will neither gain victories over prejudice and error, nor extend the empire of science. A mere theory is in truth destitute of merit of every kind, except that of a warm and misguided imagination. It demonstrates neither patience of investigation, nor rich resources of skill, nor vigorous habits of attention, nor powers of abstracting and comparing, nor extensive acquaintance with nature.

Even the careful description of his interference experiment by Young did not sway the reviewer's opinion. The latter was finally put in the position of denying that the experiment was ever done as Young described it. Scientific dogmatism could scarcely go further! The articles in the *Edinburgh Review* were not signed, but there was every reason to believe that they were written by the clever young Scottish jurist Henry Brougham, some of whose early published mathematical work had been criticized by Young in somewhat harsh terms. At any rate, Young assumed that Brougham was his assailant and Brougham apparently never denied the accusation. Brougham later moved to England and became a political celebrity, at one time (around 1830) occupying the post of Lord Chancellor. He continued his interest in science and mathematics as an avocation, and was evidently well read. However, historians of science generally agree that his attack on Young's advocacy of the wave theory of light with its dogmatic adherence to the Newtonian theory retarded the general acceptance of the wave theory by many years. It seems clear that his dogmatism was not confined to a blind defense of Newton's ideas, but also involved what we should now call an utter misunderstanding of the fruitful method of scientific explanation through bold and imaginative theorizing.

Of course, the dogmatism of the adherents of the Newtonian point did not prevail. The shoe was on the other foot when at the end of the nineteenth century the success of the wave theory was challenged by new experience which seemed to demand a particle explanation. Vigorous controversy ensued. Fortunately, dogmatism did not prevent the development of a new and successful synthesis.

Another celebrated illustration of dogmatism in nineteenth-century science is the attitude of Lord Kelvin toward physical explanation. He was so impressed with the success of mechanics as a physical theory that he managed to convince himself that all physical experience must be describable in mechanical terms. Hence, when Maxwell created the electromagnetic theory of light Kelvin announced that he could not understand it unless it were ultimately reduced to mechanical terms. Though the theory was first announced in 1864, Kelvin, to the time of his death in 1907, never became reconciled to it and indeed argued vigorously against it on numerous occasions. It is perfectly true that his opposition was obviously not motivated merely by blind dogmatism such as that shown by Brougham in the case of Young. Kelvin was a great scientist and a thoroughly honest one. But he had made up his mind that a theory which did not involve a picturable model (characteristic of mechanical descriptions) was not understandable. Now Maxwell in his introduction of the concept of displacement current, without which he could never have predicted the existence of electromagnetic waves, broke away from this tradition of the model. It was a bold stroke and set an example which twentieth-century physicists have not been slow to follow. Many of the late nineteenth-century physicists were undoubtedly convinced of the value of Maxwell's results when Hertz, in 1886, verified his predictions by the experimental discovery of the properties of electromagnetic waves. But this did not satisfy Kelvin; he wanted a mechanical model at the basis of the theory before he could be happy about it. His dogmatism did have one wholesome result: it focused attention on one fundamental weakness in the working out of Maxwell's theory, namely, in the interaction between light waves and the atoms of which the atomic theory assumes all substances to consist. This also worried Maxwell, as it has worried many physicists since; it is a problem which has only comparatively recently received a solution.

After a careful consideration of the preceding illustrations, the reader may object that any scientific theory may become the basis for the dogmatic adherence to a particular point of view, and this is indeed the case. However, certain principles embracing large regions of scientific thought offer peculiarly tempting opportunities to the enthusiast. A good illustration is provided by the so-called "principles of renunciation or impotence," which state in categorical fashion the inability of human beings ever to carry out certain processes. The late Sir Edmund Whittaker called particular attention to these in his presidential address to the Royal Society of Edinburgh in October, 1941, and emphasized what he considered to be their great value in scientific explanation. There are numerous examples of principles which can be interpreted from this point of view. The indeterminacy principle with its generalization to the principle of complementarity, considered in the previous section, is a case in point. It can be interpreted to mean that we must forever renounce the possibility of simultaneously measuring the position and momentum of a particle with complete precision. Another example is provided by the second law of thermodynamics in the formulation which says that it is impossible for any self-acting machine continuously to convey heat from one body to another of higher temperature. This certainly conveys a categorical expression of impotence and implies an irrevocable renunciation of any attempt to carry out a process which the principle proclaims to be impossible. Still another example comes from the theory of relativity, in accordance with the basic principle of which it is impossible to detect absolute motion of an inertial system.

Now statements of this kind certainly have a dogmatic air about them. For this reason they are probably psychologically unfortunate for the advancement of science. When you tell a person that there are certain things which are forever impossible for him to do, you are apt to arouse in him a spirit of hostile opposition not conducive to a dispassionate examination of the circumstances surrounding the choice of such a postulate. Fortunately, the kind of dogmatism involved in renunciation principles is rather readily overcome by phrasing the principles in positive rather than negative form. This can always be done. Thus, the second law of thermodynamics can be phrased in the form: "the entropy of the universe tends to increase." Statistically interpreted (as is the prevalent tendency among physicists), this avoids the dogmatic tone associated with the renunciatory mode of expression, for it recognizes possible fluctuations in the actual behavior of the entropy; its increase is only a statistical average.[7]

We could present many other illustrations of incipient dogmatism in science and its possible dangers for the future development of science—but, as we have suggested, these dangers are somewhat mitigated by the fact that in a free society scientists are able to criticize each other's views with complete candor. This prevents the hardening of scientific theories into long-lasting dogmas. The situation is unfortunately otherwise when scientists must function in a state which subscribes to a particular philosophy and insists that all science conform to the dictates of this philosophy. The Soviet Union now provides an example of this. The official philosophy of the Communist Party which controls the state is dialectical materialism and the government endeavors to force scientific investigations into conformity with it. The result is not science but scientism[8] and has undoubtedly impeded the progress of Russian science, notably along biological lines. That the effect has not been so noticeable in the physical sciences is undoubtedly due to their more abstract and mathematical character and the possibility of paying lip service to the official philosophy without letting it interfere with actual serious investigations; in other words, pulling the wool over the eyes of the Party. Nevertheless, the whole situation is not wholesome when one considers the complete freedom which genuine science needs for its development. We shall return to this problem in a later chapter in connection with the relations between science and the state. We content ourselves here with the reflection that science does itself a disservice when it ties itself to a dogmatic philosophy, and it suffers when through an external agency it is forced into such an association.

[7] R. B. Lindsay, "Impotence Principles in Modern Physics," *Scientific Monthly*, *67*, 50 (1948).

[8] Cf. Gustav Wetter, "Ideology and Science in the Soviet Union. Recent Developments," *Daedalus* (Journal of the American Academy of Arts and Sciences [Summer, 1960]) , p. 581.

CHAPTER 5

SCIENCE AND

HISTORY

The Nature of History. Definitions and Theories

What is history? The reader may amuse himself on a rainy Sunday afternoon by looking up what is said about the subject in a dictionary of quotations. Evidently the contemplation of it has moved distinguished people in various ways and with results expressed with various degrees of epigrammatic emphasis. Thus Viscount Bolingbroke quotes Dionysius of Halicarnassus as responsible for the famous statement: "History is philosophy teaching by example." On the other hand, Georg Wilhelm Hegel felt constrained to confess that "What experience and history teach is this—that people and governments never have learned anything from history, or acted on principles deduced from it." Thomas Carlyle, who thought a good deal about history and devoted much of his life to writing it, was convinced that "the history of the world is but the biography of great men." The epigram attributed to Sir John Robert Seeley, "History is past politics, and politics present history," has appealed to many impressed with the importance of government in the affairs of men, and, in turn, has driven many away from the study of history since they have found it unutterably boring.

Some look upon history as a fascinating record of human successes and failures, while others agree with Henry Ford, who said "History is bunk." In somewhat similar vein is the assertion that history is a "lie agreed upon." So both the enthusiasts and the cynics have their say. As we have remarked previously, such dogmatic statements have little value except to emphasize personal prejudices.

Sometimes an opinion about history is uttered almost unconsciously along with a view about intellectual activities in general. The following story is told of the historian Edward Gibbon, who was writing his famous *Decline and Fall of the Roman Empire* about the time of the American Revolution. Just after the publication of the second volume of this work he happened to meet the Duke of Gloucester, who referred to the event in these terms: "What! Another damned thick book! Nothing but scribble, scribble, scribble, eh, Mr. Gibbon!" His Grace could hardly have expressed his view on history more felicitously had he written a book on the subject.

For the purposes of this chapter we shall be satisfied with a more sober appraisal and agree with W. H. Walsh[1] that the historian "aims at an intelligent reconstruction of the past." Even this modest description evidently needs some interpretation. Thus the question immediately arises, what past is meant? Is it the past of the cosmos or the earth considered as physical systems? Or is it the past of living organisms, evolving from simple forms to the more complicated creatures now inhabiting the earth? Or should it refer only to the past of human beings? Strictly speaking, any one of these interpretations would satisfy Walsh's description. Actually, it is well known that professional historians have in general confined their attention to the human past and have left to others the history of nature. However, it is of some significance that in recent times, history departments in universities are showing favor to historians of science, who try to trace the evolution of scientific concepts, their technological applications as well as their impact on society. But as a rule each intellectual discipline, art, music, engineering, etc., has been left responsible for its own history, and the scientists, in particular the astrophysicists, have the job of constructing the past of the universe. It would appear to be an ideal of the future that the "past" for history could become the whole past of human experience in the widest possible sense.

The next question is, what shall we mean by "reconstruction"? According to the great German historian of the nineteenth century, Leopold von Ranke, the historian should confine his attention to "precisely what happened"; that is, the plain, unvarnished facts

[1] *Philosophy of History* (New York: Harper Torchbooks, Harper & Brothers, 1960), p. 29.

should be presented with faithful adherence to the available documentation. This procedure can make history a pretty dull narrative indeed, and many contemporary historical works exemplify this fully. A purely descriptive narration of a succession of events with no attempt to see relations among them might well be as meaningless as the description of a disconnected series of physical phenomena. At any rate, Walsh by the interpolation of the word "intelligent" clearly implies that the narrative, to be genuine history, must involve some sort of interpretation or the endeavor to find meaning in the events described. Perhaps the word "intelligible" is more adequate for the purpose. Just as scientists have not been content merely with the observation of "how things go," but are continually goaded by some inner desire to "understand," so most historians wish to understand the record. By this they presumably mean the providing of an explanation of why events took the course they did, i.e., *why* things happened as the record indicates. This would seem to imply a close connection with the method of the scientist in creating theories to explain human experience. From this point of view history must also operate in terms of theories. Thus, it is not sufficient for the historian to catalogue as accurately as possible the principal events in the Roman world from A.D. 17 to A.D. 410; he must also provide a theory for the steady decay of that civilization which is commonly referred to as the decline and fall of the Roman Empire. Similarly, the American Revolution cannot be considered as understood without a theory of its origin and purpose. It hardly takes a profound knowledge of history to grasp at once the enormous difficulties associated with such a program. The mere problem of what constitutes a civilization and the meaning of its decay involves matters of definition on which historians of equal ability and insight can obviously maintain quite different points of view—and often actually do.

Is There a Science of History?

It is now time to become more specific and categorical about the relations between science and history and ask: is history a science or can there be a science of history? Some historians, notably J. B. Bury, for example, have insisted that history is a science. If this is indeed the case, it ought to be possible to find in the activities of the historian the elements of the method of science set forth in Chapter 2. Can we really say that history is also a method for the description, creation and understanding of human experience? And do the terms description, creation and understanding mean essentially the same thing for history as for natural science?

Let us look at description first. In science this means the search for regularities or routines of experience, expressed in terms of scientific laws. Does the historian look for such regularities in past human behavior? The answer would appear to be that some do while others do not. Edward Gibbon devoted himself assiduously to an examination of the causes of the decline and fall of the Roman Empire, but it is nowhere evident that he made a thoroughgoing study of any regularities evident in the growth and decay of imperial systems in general. Carlyle wrote of the French Revolution, but provided no recipe for the making of revolutions in general.

On the other hand, Arnold Toynbee in his *Study of History* has endeavored to trace the cycles of rise and decline in a whole series of civilizations and has developed a theoretical point of view by which he essays to use his studies to predict the future. Similarly, Oswald Spengler has constructed a theory of history from which he foretells with considerable assurance and presumably some satisfaction the collapse of Western civilization.

In any discussion of the relations between history and science, the name of Henry Thomas Buckle immediately comes to mind. In his major work, *History of Civilization in England*, Buckle endeavored to support the thesis that history is indeed a science in the sense that it is the account of human actions, and that such actions manifest on the average regularities of the same intrinsic kind as those exhibited by the phenomena of so-called inorganic nature. It is then the task of the historian to seek out such regularities: there can be, says Buckle, no real history unless this obligation is faced and discharged. To those who would emphasize the utter hopelessness of the task because of the complicated mental and physical make-up of man and the almost infinite complexity of his relations with his fellows, Buckle points out the statistical regularities which L. H. J. Quetelet (the famous Belgian astronomer and statistician) had turned up in such things as suicide rates and marriage rates, etc., in

Western European countries. His suggestion is that historians should study these regularities in the light of the reciprocal relations between man and his natural environment.

Though Buckle makes out a good case for the importance of the effect of the physical environment on the actions of men as well as the influence of their ideas, e.g., the use of skepticism in inquiry, it does not appear that he succeeds in really setting up historical theories from which future events can be predicted. The latter part of his work largely reduces to a recital of events from the point of view of those preconceptions of whose fundamental validity he has persuaded himself. He was indeed convinced that all the actions of men can be understood in terms of a "vast scheme of universal order," not understood to be sure in his time but which the rapid acceleration in the process of inquiry would eventually substantiate. He predicted that by the middle of the twentieth century this attitude would become commonplace among historians. Unfortunately for his thesis, the fact is quite otherwise! Historians are still arguing this point and there appears little likelihood the matter will be settled on Buckle's terms or even be settled at all as long as the thinking processes of human beings remotely resemble those now in evidence.

Among modern historians, one finds a rather sharp cleavage between those who, like H. A. L. Fisher,[2] insist that history is completely patternless, that is, a record of something which just happened and has no significance for the future, and those who, like E. H. Carr,[3] feel strongly that there has to be a *pattern* in history, but that it is imposed on the events by the historian and is therefore colored by his prejudices. His pattern may be that of a steady progress toward higher civilization in the sense of Bury, Condorcet and Montesquieu, or it may be that of an irrevocable decay, as Spengler and others like him have viewed things.

Though Carr believes that history without pattern is without value and indeed almost meaningless, he does not thereby draw a strong analogy between history and science. In fact, to him, such an analogy is false: in science we have effects forever repeating themselves on the emergence of the same causes; there is no memory of

[2] *History of Europe*, 3 vols. (Boston: Houghton Mifflin Co., 1935).
[3] *The New Society* (London: Macmillan, 1953).

earlier experience and no carry-over. In history, on the other hand, every time something happens, it does so under a new set of circumstances, for everything relating to human beings is in a continual process of change—we are never able at any time or any place to reproduce the conditions prevailing at an earlier time; hence, repetition, in the sense of science, is impossible.

Now this is an effective presentation of a plausible point of view, which has had much currency among nonscientists. Unfortunately, it suffers from distortion, based on an inadequate interpretation of science. The ability to locate routines or patterns of experience, so characteristic of successful scientific endeavor, is largely due, as we have seen in Chapter 2, to the willingness to abstract from the totality of experience small domains for intensive examination, with the hope that what goes on in these regions will not be seriously interfered with by the extraneous environment which is being neglected. This means in technical terms that science deals essentially with *isolated systems*. Obviously, these are highly idealized, for our experience never actually comes to us in such neat packages. Nevertheless, the success of science means that we do find in experience something *approximating* the ideal systems invented in scientific theorizing, though we can be fairly sure that in no repetition of a scientific experiment, for example, do all external conditions ever remain precisely the same as on previous occasions. To say that one cannot in principle ascertain similar patterns or regularities in abstracted domains of human experience as recorded historically is simply begging the question. Perhaps they have not been looked for with enough insight. Actually, it has taken the human race a long time to build the effective method we now call science in the face of what looked, and still looks to the not too careful observer like a chaotic flux of sensation in which change seemed to be the only sure element. There would seem to be some hope for history after all! And the attempt to see an utter difference between the two disciplines might turn out to be an unnecessary counsel of despair.

One can indeed muster an even stronger argument to cast doubt on Carr's view. Science and particularly physical science has not confined its attention to isolated systems in equilibrium, for which it has proved possible to build theories of relative simplicity. It has not hesitated to tackle the more difficult problem of following systems moving through nonequilibrium states, as, for example, a gas expanding freely from a high-pressure region into one at lower pressure. Such a process is termed irreversible and might be thought to be outside the scope of scientific description, somewhat as a sweeping social phenomenon like a revolution might seem to defy any rational understanding. But by the judicious use of statistical reasoning, much progress has been made in the understanding of irreversible processes. To take another illustration, it has proved possible in physics to handle cases where the behavior of a system depends not on the influences instantaneously exerted on it but on its whole past history. Thus, when a solid like a wire is stressed, the corresponding deformation in general depends on the previous stress history. To handle such problems, a special technique known as hereditary mechanics has been devised. It demands a rather high-powered branch of mathematics, namely, integral equations, for its successful exploitation, but the point is that it represents an effective attack on what may properly be called a historical problem in science.

But if history is after all a science, it should be possible to develop historical theories which predict regularities in the course of history, and hence can serve as a basis for predicting the future behavior of man in society. The thesis that this is possible has been called *historicism*.[4] It looks upon history as the theory at the basis of all sociological activity, i.e., it gives the direction of social advance. In much that we have said earlier in this section, we have been implicitly dealing with the historicist thesis. It must not be supposed that historicism implies that because it finds regularities in historical development it therefore endorses a complete analogy between science as developed by the natural scientist (the kind of science we discussed in Chapter 2) and history. Most historicists deny this vehemently, taking the stand that the analytical features of orthodox science with its emphasis on the employment of abstract mathematical symbolism are wholly inadequate to handle social phenomena, in which appropriate concepts are notoriously difficult to pin down as precisely as, for example, they are in physics. Moreover, the his-

---

[4] Cf. Karl R. Popper, *The Poverty of Historicism* (London: Routledge and Kegan Paul, 1957), and Geoffrey Barraclough, *History in a Changing World* (Oxford: Basil Blackwell, 1955). (The latter contains numerous references to the cult of historicism.)

toricist is apt to view the world in the light of purpose, which he takes for granted is utterly foreign to the scientist with his causal or at any rate deterministic attitude. That he thereby betrays an ignorance of the presence of teleological factors in certain scientific theories is clear from some of the discussion in Chapter 2, where we pointed out the teleological import of a principle like that of Hamilton's in the theory of mechanics.

Another vital difference between historicism and science has been sought in the essentially holistic attitude of the former. This is the view that the only proper way to deal with the systems which are revealed to our experience is to treat them as wholes and not as aggregates of parts. It is the antithesis of the atomic hypothesis, according to which the properties of a system are due to the summation of the behavior of its constituents. A group of individuals, for example, will display characteristics which depend on the whole previous history of the group and these cannot be predicted simply on the basis of a knowledge of the behavior of the individuals themselves. The holistic point of view has had a strong appeal to biologists, who must handle and somehow account for the actions of organisms, and find it difficult to do this in terms simply of an analysis of the structure and components of the organism.

Karl R. Popper[5] attacks the whole concept of holism as ordinarily understood by emphasizing that in the strict sense one can never hope to study any elaborate system as a whole. All one does is to concentrate on one aspect. The larger the system, the more meaningless it is to try to assert anything about its properties as a whole. According to this view, the thoroughgoing holists are merely deluding themselves. This is a plausible criticism which will probably appeal to most scientists until they reflect on the way in which thermodynamics concerns itself with the macroscopic properties of large-scale systems in terms of such concepts as energy, entropy and enthalpy. Admittedly, these concepts relate only to certain specified properties and by no means encompass everything that one can say about the system, and thus can hardly be considered as holistic in the absolute sense; but they are, so to speak, relatively more holistic in character than the atomic and molecular concepts used to describe the system in statistical mechanics. To this extent, then, the holistic idea enters even into physical science.

[5] *Op. cit.,* pp. 26 ff.

But this does not get us very far in support of historicism: we mention it only to be fair. Popper claims finally to dispose completely of the validity of the historicist's point of view by a logical argument. For this purpose he effectively equates historicism to the attempt to predict the future course of history. He takes for granted that everyone will agree that the growth of human knowledge affects the course of history. This seems plausible—for the moment, let us accept it. But no sensible person ventures to predict the future development of human knowledge. It therefore follows that the prediction of human history is impossible. Popper believes that the argument can be put into logically unassailable form.

Let us look again at the major premise of this deceptively simple syllogism. The very influence of science on civilization through technology would appear to guarantee the validity of this assumption. Certainly, the external modes of living in the twentieth century are very different from what the historical records tell us of life in the fifteenth century, and we justifiably attribute this to the change in technology brought about by increase in human knowledge of the natural environment. Nor must we overlook the philosophical influence of the changing concepts of science on human thinking, which we have already emphasized as having much to do with the way people in general look at their experiences.

The great question, however, will forever remain: Are the influences of science decisive? If they are, we can hardly avoid Popper's conclusion, for the second part of his syllogism appears unassailable. The future course of science is so thoroughly dependent on the imaginative powers of individual men that any attempt to predict it seems illusory. Is it possible that though the influence of science on society as a whole is very great indeed there still remain fundamental drives inherent in man as an animal which will not be tamed by science, at least in the foreseeable future, and which are more decisive for his ultimate destiny? Recall, in this connection, Sir Charles G. Darwin's interesting thesis in his book *The Next Million Years*[6] that man is the only wild animal left on earth and if he is to be tamed it must be only by one of his own kind. But that still leaves one untamed animal around by definition, and it is the very nature of the wild animal that his behavior is unpredictable.

We need not take these latter considerations too seriously to feel

[6] (London: Rupert Hart—Davis, 1952), Chapter 7.

fairly sure that even if the influence of science on civilization is not decisive and the future course of history depends on the behavior of a master mind who succeeds in taming his fellows, we are still faced by a failure of historicism to make its case in any thorough-going fashion.

Here we shall let the matter rest, merely observing that the exist-ence of historical theories and derived laws appears to be of doubtful value. But this certainly does not rule out the pursuit of historical studies in a manner analogous to that used by the natural scientist. This suggests that we should examine other possible relations be-tween science and history. One of these is clearly the problem of objectivity. It deserves a section by itself.

Objectivity in Science and History

One of the obvious characteristics of science, at any rate in the popular view, is its objectivity. This means, as we took occasion to point out in Chapter 3, where we looked into the relation between science and the humanities, that in the mode of describing experi-ence which we term scientific, it is possible to secure agreement among competent observers as to what is being observed as well as to an acceptable description of the same. Thus, when a physicist decides to describe the relation between the pressure and volume of a gas at constant temperature, and by a long series of careful ex-perimental measurements comes up with the law which we call Boyle's law (Chapter 2), it is not his mere subjective reaction to the phenomena which is in question. He can tell other scientists what he has done and what he has found; it then becomes possible for others to repeat his work and hopefully to duplicate his results. At any rate this has happened often enough to justify the conclusion that there is a high degree of objectivity in the discipline; the agree-ment among a host of careful observers and thinkers seems to justify the assumption that what they are working with and talking about is outside themselves and is common experience to them and indeed to all who would take the trouble to follow the same operational procedure.

If history is to resemble science it is natural to expect that it will in some way exemplify this property of objectivity. Thus, when his-torians confront the same set of facts they should presumably come up with the same description and the same interpretation if they are to operate scientifically. But it is notorious that they do not. The history of England written by David Hume is a different story from the history by James Anthony Froude. George Bancroft, in general, gave a different interpretation to many phases of American history from that presented by, say, S. E. Morison and H. S. Commager. By and large, these historians had the same access to the relevant state papers. But they brought to their task different points of view based on a great variety of factors—education, cultural environment, temperament, etc. Even more extreme is the difference between such a book as *A History of the Warfare of Science with Theology in Christendom,* by Andrew Dickson White, and the orthodox histories of the "creation" and the "fall of man." It seems, indeed, to be a rather melancholy fact that positive views on religion have a peculiarly strong and often unfortunate influence on the writing of history. The controversy over the treatment accorded by Edward Gibbon to the rise of Christianity in his *Decline and Fall of the Roman Empire* is an example. His assailants claimed that his freethinking proclivities made him irreverent and cynical. He retorted that he was merely being faithful to the facts as he found them recorded. Perhaps he was deceiving himself. The fact remains that the theological historians managed to tell a different story.

All this of course raises the gnawing doubt that there can really be any genuine objectivity in history, if each historian is privileged to introduce his own preconceptions and prejudices into his interpretation of what happened. And yet, what else can we really expect? A history which has had long a reputation for objectivity and fidelity to the facts is Julius Caesar's *De Bello Gallico.* The style is simple and direct, and the narrative persuasive to such an extent that no one (certainly not the untold generations of schoolboys who have been exposed to it) can see much reason to doubt it. Caesar tells us candidly how the Romans and Gauls drew up their respective battle ranks, what he directed Labienus to do, how the cavalry behaved, how the enemy was finally routed, what the casualties were on both sides and how the peace-seeking emissaries were treated, etc. One can almost imagine oneself there and seeing it all as an eyewitness! But then, how do we know that the famous author is telling the truth? Practically all the available sources for this period of history in Gaul rely on Caesar's account, though certain episodes are presented with minor variations, as by Dio Cassius, for example. We know that Gaul became Romanized and perhaps this is evidence enough for the validity of Caesar's story. And yet, his history of the civil war in which he made himself master of Rome has been thought by some to contain some questionable assertions designed to put the author in a more favorable light. The whole career of Caesar has in fact aroused considerable controversy and difference of opinion among recognized historians, and thus provides another illustration of the difficulty the scientist has in finding something akin to his own idea of objectivity in the operations of historians.

If indeed historians are permitted to interpret the record of the past in the light of their own prejudices, how can there possibly be any objectivity in history? It might then seem that the matter could be dismissed without further consideration. But W. H. Walsh feels the problem deserves a less casual treatment. He argues persuasively that before we reach the conclusion that the writing of history is purely subjective, i.e., what we may call historical skepticism, we ought to consider two other possible points of view. He calls the first of these the "perspective theory" and the second the "theory of an objective historical consciousness."

In brief, the perspective theory holds, though it is inevitable that each historian will approach his reconstruction of the past from the standpoint of his own preconceptions and built-in prejudices, his account, within this framework, can still be objective if he is honest enough to examine his evidence and his sources with faithful adherence to those standards which all careful observers of phenomena of any kind accept, and if he follows the common rules of logic so as to avoid gross fallacies of reasoning. In this view, it is possible, for instance, for both the freethinker and the devoutly religious historian to be equally objective and at the same time draw different conclusions from the same set of facts. An appropriate illustration is found in Edward Gibbon's *Decline and Fall of the Roman Empire,* especially in the well-known Chapters XV and XVI, in which the author discusses the rise of Christianity.[7] Gibbon was severely criti-

[7] *The History of the Decline and Fall of the Roman Empire,* with notes by H. H. Milman, two vols. (n.d.) (London: Ward, Lock and Co.) There are many editions of this famous work. The one here cited is especially appropriate for our purpose because of the editing by Dean Milman, the eminent Anglican divine.

cized for his allegedly unfair treatment of the development of the Christian religion during the early days of the empire, but it is interesting to note the view of a celebrated theologian like Milman on this point. Milman makes clear that he has no quarrel wtih Gibbon over the facts he sets forth concerning the practice of the early Christians. Here, at any rate, the distinguished historian is admitted to be objective and accurate. What Milman objects to is essentially the unwillingness of Gibbon to stress the divine origin of the religion as the essential reason for its successful spread, rather than the more prosaic explanations adduced by the historian, e.g., the intolerant zeal of the Christians, their belief in a future life, their pure and austere morals, their cohesiveness and discipline, etc., all of which got them into trouble with the Roman authorities but led ultimately to their thriving on persecution. What it comes down to is that Gibbon was objective in his adherence to the facts as he found them recorded in the sources available to him, but he failed to display the reverence toward the theological implications of the facts which his theological adversaries demanded of any historian of Christianity. Here we have the combination of objectivity with subjectivity which Walsh evidently associates with the perspective theory of history.

We must not overlook one serious handicap to which every historian is subject: the provisional character of his sources. Historiography is not static, and new evidence of man's past is being continually brought to light by the labors of archeologists, searchers through dusty and neglected archives, discoverers of forgotten manuscripts in national libraries as well as in private hands. Much of this activity is indeed being facilitated by scientific development. Two examples are provided by the method of dating ancient organic materials by the study of their content of the radioactive carbon 14 isotope, and by the use of special chemical and mechanical means for unrolling ancient scrolls, like those discovered in 1956 in caves near the Dead Sea. Dating by the carbon 14 method is possible only for objects not much more than 20,000 years old, but trials with other radioactive isotopes, notably rhenium 187 and iodine 129, have suggested to W. F. Libby, the inventor of the carbon 14 method, that dating of fossils many millions of years old may turn out to be possible. This may have an important bearing on the human record also.

It is clear that historical objectivity must be limited to the record as it stands during the investigations of each particular historian, just as his subjective interpretation is also a function of his own time and environment. In this respect, the situation is by no means so different from that in science as might at first be thought. The scientific method is continually creating new experience and objectivity in science certainly changes from age to age. Moreover, theories, which constitute the subjective aspect of science, are also closely conditioned by the whole cultural milieu of the scientist, as we have already seen.

The final suggestion of Walsh concerning objectivity in history is that it *may* ultimately be possible to achieve a genuine "objective historical consciousness." By this, he presumably means that by the honest and perceptive endeavors of generations of historians agreement may gradually be reached on a set of basic presuppositions which will serve as the touchstone for all historical investigation. In other words, a single all-embracing theory of history may emerge in spite of all the difficulties which have been cited. This is something like assuming that with the progress of science, a uniform point of view with respect to man's experience as a whole will ultimately come into existence and all persons will accept it. From the conception of science presented in this book, the likelihood of this outcome is very remote indeed and the same should probably be said for the grand historical synthesis. Walsh's many critical comments of the thesis in question would seem to imply that he is not very confident of it himself.

How shall we sum up the relation between objectivity in science and history? So far as the facts of the historical record are concerned, the scientist is constrained to conclude that there should be as much objectivity in history as in any branch of science. Granted that the facts are often hard to establish; but so are in general the facts about natural phenomena which are the scientist's stock in trade. When the interpretation of the facts engages the attention, subjectivity creeps in and doubtless more so in history than in science, presumably because historians are human beings dealing with knowledge about human beings in groups of all sizes, whereas scientists are able to abstract from the totality of experience to their hearts' content. But it must not be forgotten that even science has its subjective side in the imaginative invention of theories. The fact that the success of such theories impels their acceptance by large numbers of thoughtful people confers a degree of objectivity on them which intrinsically they may not deserve. And we know that fashions in scientific theories do change from age to age. Probably the strongest claim to objectivity made in the name of science is associated with the technological applications which have such an immediate impact on all people that it is hard to see anything subjective about them. History offers little which is analogous to this save political and sociological experimentation, which is notoriously difficult to carry out and assess. We shall have more to say about this later in connection with the problem of the social studies in general.

The History of Science and Its Importance for an Understanding of Modern Science and Technology

It would be inexcusable to leave off our discussion of the relation between science and history without touching on what is probably the most important point in the whole matter, namely, the history of science itself. The very influence of science on civilization which is the theme of this book dictates the necessity of paying attention to the story of science's development as a significant part of human history as a whole. We recall the saying attributed to Francis Bacon: "The history of the world without the history of science is like the Cyclops Polyphemus without his eye." Professional historians have not as a rule taken this seriously, and until comparatively recent times concern for the history of science was manifested mainly by scientists themselves or philosophers, though there have been conspicuous exceptions. With the modern concern for science and its applications, this picture is changing and the history of science is more and more being recognized in university departments of history as a significant part of the whole discipline.

The argument that the professional historian should concern himself seriously with the history of science is a simple and obvious one. If science plays the role in modern civilization that it appears to play, it becomes imperative to ascertain what its role was in the past. The fact that its influence was neglected by historical scholars does not mean that its role was negligible, but merely perhaps that scholars in their ignorance or indifference simply paid no attention to it. They were so busy concerning themselves with political history that they completely overlooked the overwhelming significance of a discovery like that of Michael Faraday's in 1831, of electromagnetic induction which ultimately led to the founding of electrical engineering with its profound effects on industrialization in the late nineteenth and early twentieth centuries. Of course, it may be argued that it is often difficult to trace early technological advances back to their pure science roots, but, if history means anything, that is precisely what the historian has an obligation to do. Hence the importance of the rise of history of science as a professional discipline.

One obvious reason for the earlier neglect of the history of science in the standard histories is that it is a very difficult discipline. In addition to the ordinary demands which history as a profession lays upon its devotees, i.e., ability to locate, read and understand primary sources of information about happenings in the past, the history of science involves understanding of what science is all about, that is, a real appreciation of modern developments. This has hitherto proved to be an endowment rarely found among professional historians.

But the history of science is fraught with an even more serious difficulty, which is indeed inherent in all historical studies but assumes a particularly formidable character when early science is involved. Assuming that the sources exist in which the ancient scientist sets forth his ideas about natural phenomena and that these can be read by competent authorities, there remains the problem of *understanding* what the material really means in terms of modern ideas. But it may be objected that the historian should merely set forth the facts as he finds them and not try to force them into any modern mold. It requires but a moment's thought to realize that this would make the history of science meaningless. To produce any understanding of the record left by an earlier scientist, the historian must translate his terminology into terms that correspond to modern nomenclature. But this at once implies the assumption of a glossary. It must be emphasized that it is not merely a problem of translating from a foreign ancient language, e.g., Latin or Greek, into modern English, though this always presents a certain amount of difficulty; the problem is one of deciding on the meaning to be given to a particular term in the original record that will convey adequately what the early writer had in mind. It is of course a difficulty associated with history of any kind, but applies with especial severity to the history of science because of the technical character of scientific language in any age. It puts a heavy obligation on the historian of science to make himself master of the whole scientific and cultural environment of the period to which he gives his attention, so that so far as possible he can think about things as scientists in that period and environment actually thought. In a strictly logical sense, what we are asking here is actually impossible, though great historical scholars have managed to approximate the conditions moderately well, or so it is believed by their contemporaries and successors. Any one who feels that this point is exaggerated should try for himself to examine an ancient scientific work such as the *De Subtilitate Rerum* of the Italian Renaissance thinker Jerome Cardan, or even the original Italian version of Galileo's famous treatise on mechanics: *Discorsi e Dimonstrazioni Matematiche intorno a due nuove scienze Attenenti alla Mecanica & Movementi Locali* (Leiden, 1638). For that matter, he will find his hands full of trouble with the Latin version of Newton's *Principia Mathematica* (1687), or with d'Alembert's *Traité de Dynamique* (1743). Even to one well versed in modern science, the words look woefully unfamiliar.

There is indeed one obvious source of help to the historian of science in examining earlier treatises. This is the fact that, in mathematical sciences at any rate, our predecessors used figures to illustrate their ideas and trains of thought. A figure is a universal language, transcending the boundaries of time and the vagaries of the written word. In science it conveys information in the same way a painting carries a message in art. It is, and presumably will be for a long time to come, one of the most valuable of pedagogic devices, and both contemporary scientist and historian of science lean on it heavily.

But it is clear in any case that in order to make progress in the history of science fundamental assumptions have to be made by the historian as to what the earlier scientist means by what he says. To all intents and purposes, this implies the construction of a theory which in logical structure is completely analogous to any scientific theory, as discussed in Chapter 2. An obviously basic hypothesis is that the author of the work under examination was trying to describe experience in a fashion bearing enough resemblance to what scientists do today to warrant the historian in considering the material scientific in character. Another typical hypothesis would be to the effect that if the early scientist uses symbolic analysis, his usage agrees with some recognizable branch of mathematics, or is at any rate in conformity with standard logic. Again, some definite operational meaning must be associated with what appear to be the basic concepts employed. Thus, if, for example, an important term is "pondus," it is imperative that it be decided whether the meaning of *mass* or *weight* is to be attributed to it. An examination of an ancient scientific text in the light of such hypotheses then permits an interpretation which can be intelligently compared with modern scientific description of similar experience.

Let us give an illustration. In the year 1600, the English physician and natural philosopher William Gilbert published his treatise *De Magnete*, generally admitted to be the first great English work of science. This was an account of magnetic phenomena as observed up to his time and included much of his own research. Gilbert was fascinated by the dependence of the dip of the magnetic compass needle on the geographical latitude. (It will be recalled that if a magnetic needle is arranged so as to rotate about a horizontal axis perpendicular to the magnetic meridian and allowed to swing freely it will set itself at an angle to the horizontal in general different from zero. This angle is known as the *dip* or *inclination*. It is 90° at the magnetic pole and zero at the magnetic equator.) Gilbert evidently hoped that the dependence of the dip on magnetic latitude might be used to determine getographical latitude and thus be of help to navigators at sea. On page 298 of the Mottelay translation[8] there is a rather elaborate diagram from which one can read (following Gilbert's instructions) the dip for any latitude in accordance with a construction Gilbert had devised. He did not present the relation in analytical form, though in 1602 an Englishman named Blundeville published, in an appendix to a book called *The Theoriques of the Seven Planets*, a table of numerical values of the dip-latitude relation based on Gilbert's geometrical construction. This table was intended for the use of navigators.

[8] William Gilbert, *De Magnete*, translated from the Latin by P. F. Mottelay (New York: John Wiley and Sons, Inc., 1893).

In his book, Gilbert gave no clear account of how he devised his dip-latitude construction; it therefore becomes an interesting exercise in the history of science to try to determine the basis for what he did.[9] In carrying out such a research, one has to bear in mind that Gilbert could have no knowledge of the mathematical theory of magnetism developed in the eighteenth and nineteenth centuries. Hence though he may have had some qualitative understanding of the significance of a uniformly magnetized sphere of iron or steel, based on his experiments with little magnetic spheres which he made for himself in order to study the behavior of tiny iron needles on them, he could hardly be expected to work out analytically the dip-latitude relation for such a sphere (this was done by K. F. Gauss in 1839). Close examination shows that Gilbert's construction does not agree precisely with that for a uniformly magnetized sphere, though curiously enough the two agree approximately for latitudes below 30°, depart widely for latitudes between 50° and 80° and come together again in the neighborhood of 90°. One possible alternative way of approaching an understanding of Gilbert's construction is to assume that he used some sort of trigonometrical formula, a reasonable assumption since the mathematical attainments of an educated person of his age would certainly include this subject. If one acts on this hypothesis, one can indeed find a trigonometric relation which leads at once to Gilbert's diagram and also provides a basis for Blundeville's tabular values. This formula reduces for small latitudes to the corresponding one for a uniformly magnetized sphere. It is a purely formal construction, and so far as the author is aware no one has ever found a distribution of magnetism in a sphere which would produce it, though this is a problem which can be solved with enough effort. However, the effort is scarcely worthwhile, since it was evidently beyond Gilbert's powers as a Renaissance scientist, and we are sure in any case that the earth is a much more complicated magnet than Gilbert's little *terrellae*, the spheres on which he did his experiments and from which he undoubtedly obtained in purely formal fashion the clue which led to his dip-latitude construction.

As problems go in the history of science, the one we have just noted is scarcely of world-shaking significance, but it does help to

[9] Cf. R. B. Lindsay, "William Gilbert and Magnetism in 1600," *American Journal of Physics, 8,* 271 (1940).

indicate the kind of problem the historian encounters when he endeavors to probe beneath the surface of a scientific work of a bygone age.

We now face the question: what specific value attaches to an acquaintance with the history of science for intelligent people? It is probably worthwhile to divide the question into two parts, first, professional scientists themselves and second, nonscientists who are nevertheless concerned with the influence of science on society. With regard to the value of the history of science for the working scientific investigator, opinions are divided. Some feel that it has no real value whatever so far as present-day research is concerned. This view has in recent times been forcefully expressed by the chemist and educator J. B. Conant, who argues that no amount of knowledge of past scientific achievements can make a man a better research investigator.[10] He notes that as eminent an authority on the history of science as George Sarton nowhere claimed that a knowledge of history would improve a scientist's research and that its value must rest on other grounds, some of which we have already touched on in the preceding pages. Conant's view is persuasive; the really energetic research worker cannot depend on out-of-date methods, he must develop his own. His progress and success will be due to his own imaginative prowess and not to a slavish imitation of older ideas. Of course, he must be acquainted with contemporary views and the accepted status of his field at the moment, but Conant remarks this rarely extends back more than fifty years. Research is on the frontier; it does not look back but forward.

The British physicist and philosopher and historian of science Herbert Dingle takes a contrary view.[11] He feels that any scientific researcher will profit by going back to the great thinkers and investigators of the past to see if we can find out how they thought about what they were doing. As he says, the results are often very surprising, since the residuum of the work of such people as it gets embalmed in the standard treatises and textbooks often if not usually presents a rather distorted view of what they actually did and thought. His notion is that clues to the psychology of scientific in-

[10] "History in the Education of Scientists," *American Scientist*, *48*, 528 (1960).
[11] *The Scientific Adventure* (London: Sir Isaac Pitman & Sons, Inc., 1952), p. 37 ff.

vention can often be picked up by this kind of historical study. The matter will doubtless remain controversial, since no one yet understands really how clever people in any age develop their ideas and make their discoveries. But we can at any rate put the question in the following way: have working scientists as a matter of fact made studies of the work of far earlier people in the same or related fields and have they profited thereby? An honest appraisal seems to justify an affirmative answer. Illustrations abound. In more than one place, Galileo acknowledges his debt to Archimedes; and it is here not just a matter of being familiar with standard scientific literature of the sixteenth and seventeenth centuries: Galileo speaks of reading the actual works of Archimedes (in one place he says "with infinite astonishment"). It is difficult to doubt that he profited from his reading.

Thomas Young and Augustine Fresnel, who did so much to establish the wave theory of light on a plausible experimental basis, studied carefully the works of Newton and Huygens of a century or more earlier. It must be emphasized again that to study Huygens' *Traité de la Lumière* (1690) in 1800 was not like getting acquainted with established scientific doctrine as one would study it in school; the Newtonian corpuscular theory of light was the one commonly taught in England, and was indeed so firmly entrenched that it took considerable courage to challenge it, as Young did. As a matter of fact, Young was bitterly attacked by Henry Brougham for his espousal of the wave theory in the explanation of the famous two-slit interference experiments,[12] as we have noted earlier.

Still another illustration is of more recent relevance and interest. In 1924, L. de Broglie founded wave mechanics (see Chapter 3) and started the movement which culminated in the theory of quantum mechanics, which now dominates present thinking in atomic physics. In his thinking, he harked back to the famous work on the relation between mechanics and optics of Sir W. R. Hamilton (Chapter 3) and C. G. J. Jacobi, published between 1834 and 1843, whose possible relevance to the connection between wave propagation and the motion of particles had long since been forgotten, though the

[12] See Alexander Wood, *Thomas Young, Natural Philosopher* (Cambridge University Press, 1954) , p. 168 ff; also George Peacock, *Life of Thomas Young* (London: John Murray, 1855), p. 174 ff.

purely dynamical parts had become enshrined in the literature of celestial mechanics through the so-called Hamilton-Jacobi equation.[13] Here again, it was not a question of being familiar with standard treatise material available to every physicist of the twentieth century. De Broglie, as he makes clear in his semipopular work,[14] had to go back to Jacobi and Hamilton for the development of his ideas. The history of mechanics and optics was certainly of great use to him. But let us suppose he had never heard of Hamilton and Jacobi and had invented wave mechanics completely independently of any knowledge of their work. Could one honestly argue that an acquaintance with this earlier material, so closely allied to what he was trying to do and so suggestive to one who was faced with the task of trying to make quantum theory fit better into atomic mechanics, would not have been of great assistance?

This whole question is not so trivial as might at first appear, since so-called "new" discoveries are being made every year which historical research demonstrates were actually known much earlier. X-rays were really discovered long before Röntgen, for example. It is perfectly true that the effects which Röntgen decided to investigate in 1895 were ignored by earlier scientists like Crookes, but it can hardly be denied that a thoughtful person who made himself acquainted with some of the queer results obtained by competent scientists in a certain general field in a previous age might use such material to very good advantage in his own investigations. Of course, he must be able to separate the wheat from the chaff, and he must himself display imagination.

We close this section with a brief résumé of the value of the history of science in general culture. We need scarcely do more than reiterate the remarks at the beginning, that if history as a whole is an ingredient of culture in our civilization, the story of the evolution of science is an essential component and its absence necessarily leads to distortion. It is important that every educated citizen realize that modern technology came into existence not just by chance, nor by the activities of politicians, but primarily through the efforts of those

[13] Cf., for instance, Lindsay and Margenau, *Foundations of Physics* (New York: John Wiley and Sons, Inc., 1936). Paperback edition by Dover Publications, 1957, Chapter 3.

[14] L. de Broglie, *The Revolution in Physics*, trans. 1953 (New York: The Noonday Press), Chapters 1 and 8.

who have thought hard and long about natural phenomena. Some acquaintance with the lives and ideas of scientists is a part of the cultural equipment of educated people. A serious question indeed arises: can it be taught, as ordinary history is taught? Obviously, in teaching it, the educator must convey some meaning of science itself, and this is certainly more difficult than ordinary political history. Possibly the best way to do it is to combine it with the teaching of science itself. We shall revert to this in a subsequent chapter when we take up the problem of education in science.

Science and Social Studies

History is usually referred to as a social study, but shares its status in this respect with economics, sociology and political science. It therefore becomes of interest to extend the considerations of this chapter to explore the relation between science and these other social studies. Obviously, we face the same questions as previously, e.g., are the social studies sciences in the sense in which we have been using this term, or if not, do they share any of the significant aspects of science, like its objectivity, etc.?

The problem thus posed is an old one and has stirred up a lot of controversy, which still persists. Thus Auguste Comte, the French philosopher usually considered the founder, or, more accurately, the foreshadower of sociology, had no doubt about the scientific character of the discipline he believed he was creating. In his well-known hierarchy of the sciences, he put mathematics at the base because of its presumptive precision and logical unassailability (an optimistic premise on his part!) and sociology at the top, the least developed of the sciences, but the one that in a sense is the "master" science, embracing all the rest since it involves description and understanding of the social relations of human beings, who themselves create the other branches of science. Comte admitted the vagueness, qualitative character and lack of precision of sociology as a science, but felt that these drawbacks would be remedied by intensive study and the passage of time. His optimism in this respect has not always been shared by more recent thinkers, even in his own particular field. Thus Russell Kirk, the American political scientist, derides the claim of any of the social studies to be sciences in the sense of

natural science.[15] We shall examine his criticisms and others of similar nature in detail presently, but merely note in passing his attack
on the purely empirical approach of the modern sociologist, the
tendency in sociological writings toward the invention of complicated terminology without precise definition, the deficiency in imagination and the neglect of judgments in the modern study of social
relations. Kirk is convinced that the term "social science" is misleading and illusory and should be abandoned. In essential agreement
with this view was the late Charles A. Beard, the well-known American historian, who in his *The Nature of the Social Sciences* (1934)
contended that because the social disciplines cannot make predictions of the future course of society, there is no such thing as social
science in the sense of natural science.

On the other hand, the roster of those who have sought to study
social relations from the standpoint of the scientific method is long
and distinguished. We need to mention only John Stuart Mill, W.
Stanley Jevons, Lester F. Ward, Herbert Spencer, Max Weber and
Vilfredo Pareto, all profound thinkers, to see that the claim of
sociology to possess scientific aspects must be taken seriously and
not dismissed lightly.

What do sociology and the social studies in general seek to do?
They attempt to describe and presumably to understand the behavior of human beings in society, that is, as a function of their
relations to each other. The division of interest among the various
special social disciplines is largely arbitrary. Sociology proper proceeds from the fundamental assumption that since the human infant
is born in a helpless state into a social group, his whole future development is conditioned by his relation to society. Hence we cannot
really hope to understand human beings except in the social context.
From this point, the subject proceeds to consider such matters as organization, e.g., family, tribe and nation, social change, population
problems and demography. Economics, as a special social study,
concentrates attention on man in society as a producer, distributor
and consumer of wealth. Strictly speaking, this would come under
the social organization aspect of sociology, since it relates to a facet
of man's social activity which has become elaborately systematized.

[15] "Is Social Science Scientific?" *New York Times Sunday Magazine Section*,
June 25, 1961.

Those who have been attracted by the clink of the coin and the rustle of bank notes and the peculiar fascination such stuff has come to have to all mature human beings have preferred to keep their meditations apart. They have found plenty to think about, and though the coin no longer possesses the importance it once had, the economist is deeply involved in the problems inherent in the physical basis and source of all wealth, namely, energy.

Another important aspect of social organization is, of course, government. Here again, the overwhelming importance to the human race of the questions who rules and bosses whom and how is it done has justified the establishment of political science as a separate discipline.

For the purpose of our present discussion it will not be necessary to consider each social study separately in its relation to science. Their outlooks and methods are sufficiently alike so that what is said for one will do essentially for all. Generalizations about historical development are always hazardous, but it may perhaps safely be said that the general tendency of students of society in the nineteenth century was to establish a set of principles, made inductively plausible as a result of common-sense observations, and then to follow out their deductive consequences. The system builders, e.g., Spencer, Ward, Weber and Pareto, owed much to analogies with physical and biological science. The phrases "social statics" and "social dynamics" so common in the writings of such men are sufficiently familiar to make this point plain. Spencer in particular was greatly influenced by the theory of evolution and made this the central theme in his social philosophy. Elaborate quantitative attention to the facts of society was not cultivated in the sense of the natural sciences, though Quételet with his statistical studies and his disciple Buckle (already mentioned in an earlier section) are marked exceptions, showing that the empirical approach was not wholly neglected in the nineteenth century.

Herbert Spencer may be taken as reasonably typical of the system builders. He began with the basic assumption that a society is an organism; hence, in deducing social phenomena in accordance with the method of the natural sciences, he felt himself permitted (ideally, at any rate) to make use of all that biology knows about the characteristics of organisms. But Spencer knew enough biology to realize that one runs into trouble by accepting such an analogy without qualification, and so he was forced from the start to introduce exceptions. This illustrates a difficulty common to all system building in the social studies; the aggregate whose properties are to be described and explained is usually too large and too complex to permit a description in terms of a simple, all-inclusive set of assumptions. We have noted that physical science gets around this difficulty by abstracting from the totality of experience a small domain for special study—a domain that can be approximately isolated from outside influences. The early systematic sociologists did not wish to follow this pattern—they wanted a grand generalization which would account for all social phenomena. In other words, they did not depart completely from the general method of scientific explanation, but rather attempted to push it much further than the natural scientists had ever found it feasible to do.

It has been easy to criticize the sociological theorizers and system builders for their too enthusiastic elaboration of alleged analogies (isomorphisms in the language of the sophisticates) with the concepts of physical and biological science. Natural also has been the complaint that the social "laws" deduced from the assumed principles have no particular precision and are not verifiable in the sense of natural science. In fact, many critics have questioned whether there exist any real social "laws" at all. The difficulty of realizing anything resembling the isolated system of physics and the lack of ability to measure anything in the social domain have also been cited by many as rendering the efforts of the system builders illusory except as amusing little games. This and a strong emphasis on the alleged empirical character of natural science undoubtedly stimulated a revulsion from theoretical sociology and, in the early part of the twentieth century, a corresponding concern for the acquisition of more and better social "facts." By and large, it is the empiricists who dominate contemporary sociology, variously referred to as the doorbell ringers, the manufacturers of questionnaires, the statistics mongers and the dealers in standard deviations, not to mention the public opinion pollsters. Techniques for ascertaining what people on the average are thinking about important public questions will probably improve, but the present state is not very encouraging to those who are hoping for a solid empirical basis for a genuine social

science. On the other hand, statistical data on the concrete achievements of people, such as, to take a simple example, the number of
both sexes who attain degrees at an accredited institution of higher
education, are much more reliable and constitute useful social data.
At any rate, they seem useful to the natural scientist, probably because they involve numbers, and the urge to measure is irresistible
in science, as we have already pointed out in Chapter 1. It is not
hard to see that if this urge finally prevails in the social studies, social data must become statistical in character and if there are social
laws they must also be of this kind.[16]

Much of the criticism which has been leveled at the generalizations of the speculative sociologists by both natural scientists and
empirical social scientists is based on the fact that though such generalizations may pretend to have qualitative significance, they have
little or no empirical quantitative backing. An example will clarify
this point. Walter Bagehot has been considered an outstanding
nineteenth-century thinker in the social science field, particularly
with relation to political thought. His famous book *Physics and
Politics* (1869)[17] made an interesting attempt to associate the scientific developments of his time (the word "physics" in the title is of
course a misnomer—he was, strictly speaking, referring to the life
sciences and to the evolutionary notion in particular) with social and
political problems. In his second chapter, on "The Use of Conflict,"
he discusses the notion of progress and lays down what he calls
"three laws, or approximate laws," as follows:

> First. In every particular state of the world, those nations which are
> strongest tend to prevail over the others; and in certain marked peculi-
> arities the strongest tend to be best.
> Secondly. Within every particular nation the type or types of char-
> acter then and there most attractive tend to prevail; and the most
> attractive, though with exceptions, is what we call the best character.
> Thirdly. Neither of these competitions is in most historic conditions
> intensified by extrinsic forces, but in some conditions, such as those now
> prevailing in the most influential part of the world, both are so intensified.

[16] An excellent recent discussion of the statistical nature of social data and
laws is contained in Chapter 14 of Ernest Nagel, *The Structure of Science* (New
York: Harcourt, Brace and World, Inc., 1961).
[17] Walter Bagehot, *Physics and Politics*, with an introduction by Jacques Barzun
(New York: Knopf, 1948).

These generalizations, evidently considered by the author to possess the status of social laws, are indeed presented and elaborated on with more care and caution than has been true of many other writers in this field.

It is no part of our intention to discuss in detail the context in which these so-called "laws" apply, and to try to decide whether the author in the rest of his chapter has made out a good case for them. The reader who is interested may do that for himself by reading Bagehot's book. Our emphasis is of a different sort, namely, to call attention to this situation: the "laws" (or the first two of them, at any rate) purport to give a general quantitative significance to a certain social pattern. Leaving aside the fact that the pattern is vague because of the imprecision associated with the terms "strongest," "best," "marked peculiarities," etc., it is principally important to note that the use of the word "tend" at once imposes on the author an obligation to explain the magnitude of the tendency, thus ultimately implying a statistical investigation. But this is not provided, for Bagehot contents himself with calling attention to specific instances, or tacitly asks the reader to supply the evidence out of his own knowledge.

This is but a single illustration of a tendency which was typical with the systematic sociologists of the nineteenth century; their books are full of generalizations made plausible only in terms of isolated examples. This should not necessarily be construed as an adverse criticism. Perhaps this is the only way to approach the study of social phenomena so as to get competent people sufficiently interested in the subject to be willing to work at it. Certainly a natural scientist can read the pages of Bagehot with great pleasure and stimulation, even if almost every page contains statements which seem to demand more complete verification. The problem, however, is the extent to which such writings constitute a really scientific attack on the problems of man in society.

From this point of view it would seem that the empiricists are right, and the only sensible attitude to take toward sociology is to treat it as an immature science and concentrate first attention on the establishment of concrete social facts on which all impartial and competent authorities can agree; in other words, try first to guarantee objectivity in the scientific sense. On the other hand, collections of social facts without ideas about their relation can be no more scientifically meaningful than similar collections in physics. To discover or impose patterns of regularity on social observations demands the introduction of concepts; and, hence, before sociology becomes a science, some conceptualization is essential. If the example of natural science, as discussed in the earlier parts of this book, is any guide, it would seem that in sociology this process would have the greater chance of success the smaller the domains of social experience in which patterns are sought and for which concepts are constructed. In this way it should be possible to establish statistical laws relating to specific aspects of society. Actually, such laws have been discovered.

Such a law is a special case of a general law set up by G. K. Zipf.[18] The particular case relates to language, which is certainly a social phenomenon. We shall refer to it in detail in Chapter 6 when we discuss communication in general. In his endeavor to understand something of the way human beings construct language, Zipf made a statistical study of the frequency of occurrence of word tokens in a large novel in the English language (e.g., he considered James Joyce's *Ulysses,* which contains about 250,000 word tokens and about 30,000 different words or word types), and compared this frequency with the rank order. By frequency we mean simply the number of times the given word token, e.g., the article *the,* occurs in the book. By rank order we mean the number to be attached to the word type to give its rank in a list, in which the number 1 is associated with the word type used more frequently than any other, number 2 to that used next most frequently, and so on. It does not take much thought to realize that in any collection of words intended to communicate information usually considered meaningful, as the word frequency goes down, the rank order will go up. However, this gives no indication of the precise statistical relation between the two quantities. Now Zipf found that whether the material studied was a large novel or a file of an American newspaper, the relation could be expressed approximately in the form

$$P_n = A\, n^B$$

---

[18] *Human Behavior and the Principle of Least Effort* (Cambridge, Mass.: Addison-Wesley, 1949). See also Colin Cherry, *On Human Communication* (New York: Technology Press of M.I.T. and John Wiley, 1957), p. 100 ff.

where n is the integer representing the rank order (1, 2, 3 – –) and P_n is the word frequency associated with rank order n. The numbers A and B are constants. Moreover, and interestingly enough, Zipf found that B in the cases he studied comes out to be very nearly — 1. If this could be substantiated by studies of word frequency in other languages, one could then say that a regularity had been established in a social phenomenon, namely, in the human use of words. This regularity would indeed be purely statistical in character, but that would not deprive it of the status of scientific law any more than it does for statistical laws in physical science. It would seem possible to set up numerous laws of this kind in sociology and indeed some others have been observed, e.g., the relation between the population and rank order of the cities in a large country. The search for these is a scientific task and could well lead to the development of social-science theories, of limited range of applicability, to be sure, but capable of providing the same kind of understanding that physical theories give for physical laws. As a matter of fact, the attempt has been made more or less successfully to provide an understanding of Zipf's language law[19] in terms of a theory (by B. Mandelbrot) which introduces the concept of *cost* of a word or sign in a language and then assumes that language is actually constructed by people in such a way as to minimize the total cost for a certain constant rate of information. We shall examine the details in the next chapter, but need only point out here that Mandelbrot's theory actually leads to Zipf's law.

The physical scientist is perhaps naturally more impressed by the success of a theory like that of least effort as used to deduce Zipf's law than a less quantitatively-minded social scientist, who wants to see generalizations of broader scope which can cover larger domains. This is certainly his privilege, but we cannot refrain from pointing out again that the success of physical science has been due to the willingness of the physicist and chemist to abstract from the totality of experience small portions for intensive study, and indeed more due to this than to any assumed simplicity and reproducibility of his data as compared with those of social studies.

This inevitably leads us to the question: What about experimen-

[19] Cf. Colin Cherry, *op. cit.*, pp. 105, 209.

tation in social science? We have seen that this is definitely connected with the creation of experience. At the same time, in the natural sciences it serves as part, and indeed the most important part, of the method for the testing of the predictions of theories. It is commonly held and probably correctly that experiment in the sense of natural science as described in Chapter 1 is not feasible for social phenomena because of the complexity of the experience involved and the inability to control certain factors, i.e., maintain them constant, while others are being varied by determinate amounts. This however does not mean that nothing connected with the concept of experiment is possible in sociology. In his interesting book *The Tools of Social Science* (1953),[20] the British sociologist John Madge describes a number of social experiments in which reasonable control was attained. These involved changes in the behavior of groups of people when certain changes were introduced arbitrarily into their environment. The samples were small, the controls not easy, and the results perhaps left something to be desired. However, we must remember that experiment in social phenomena is not limited to operations of this kind. If the social scientist is willing to forego the type of experiment which is the genuine creation of new experience, there are many studies which can be and indeed are being made on experience already at hand in the form of statistical data to see what kind of relational patterns emerge. This is the sort of thing involved, for example, in the establishment of Zipf's language law. The field of education is a fruitful one for this kind of exploitation. But there are many other possibilities. Insurance companies thrive on the basis of the statistical law describing the relation between age and mortality. It is true that many of the relations established in this way seem only to embroider the obvious as evident to common observation. But the history of natural science seems to indicate that this is so merely because clever people have not looked deeply enough. Science is successful apparently when the appropriate concepts are developed. One somehow has to know what to look for. Then the not-so-obvious begins to emerge. This will surely happen eventually in the field of social phenomena.

We should not leave this discussion without commenting briefly

[20] John Madge, *The Tools of Social Science* (London: Longmans, Green and Co., 1953) .

on another aspect of the relation between science and social studies. This is what may be called the sociology of science. We have already emphasized in earlier chapters that the pursuit of science is itself a social undertaking. It is therefore a wholly appropriate field of study for the sociologist, who would naturally like to know the answers to such questions as: What percentage of the total population of a country may on the average be expected to go into scientific work and how does this change with time? What factors govern the number of scientists and their distribution among the various branches? How does the cultural environment affect the advancement of science? What is the effect of war on science? How may the retrieval of scientific information best be managed? All these are sociological questions of vital importance to the progress of science. We shall have occasion to touch on nearly all of them in the subsequent chapters of this book.

CHAPTER 6

SCIENCE AND

COMMUNICATION

General Considerations. Human Communication

The unabridged dictionary (Merriam Webster, 2nd ed.) defines, in the first instance, "communication" as the "act or fact of communicating," and says that to "communicate" is to "impart, bestow or convey . . . or to give by way of information." As a second meaning, "communication" becomes "intercourse by words, letters or messages; [the] interchange of thoughts or opinions."

Even without regard to the dictionary every person who deems himself intelligent believes he knows what communication is, believes himself competent to indulge in it, and is ready to admit its enormous significance in civilization. We can actually study it on two levels. In the first place, man's knowledge of the world of experience can be legitimately considered as the result of the communication of information to him from the external stimuli of his sensations, the process not reaching completion of course until something happens in the central nervous system after the sensation takes place. When we look at a star it is proper to say that by virtue of the light entering the eye information about the star is being communicated to us. In this sense, the whole of science as we are considering it in this book is based essentially on the possibility of communication to human beings. Nature is forever communicating to us, and by our experiments we are actively expanding this communication process.

On the second and more familiar level, human beings communicate with each other. Of course, it is possible that many of the lower animals do likewise, but this is a problem demanding much scientific investigation. In the case of human beings, it is in many respects the very basis of our civilization. Without it, it is hard to see how society could exist. We doubtless take it entirely too much for granted, though we are all painfully aware when for some reason it breaks down either between individuals or groups. Certainly it is a fair field for scientific investigation, and this is also appropriate since without communication there could be no science as we know it. The scientist, as a human being, simply has to tell somebody besides himself what he has done and the method of communicating is an integral part of the scientific endeavor.

The psychophysicist S. S. Stevens has given a short definition of communication[1] as "the discriminatory response of an organism to a stimulus." As Colin Cherry observes,[2] this definition probably places too much emphasis on the "response" and scarcely does sufficient justice to the *relation* involving both stimulus and response. In this sense, communication is the whole process of stimulus leading to response. Thus, in communication via ordinary speech, the stimulus is due to the sound produced by the speech organs of one person and the response is something that happens in the brain of the auditor when the sound wave strikes his ear and he is said to *hear* it. Of course, the whole process is closely tied to the physical nature of the sound emitted and received. This will be discussed below.

But speech and hearing form a relatively sophisticated mode of human communication. Preceding this by a wide time interval came communication via gestures such as grimaces, smiles, fist shaking, dancing, kicks, etc. The exchange of goods and money also constitutes a form of communication. These, though ancient, persist to this day as part of social intercourse. Communication by written signs probably came later—language in its written form followed speech.

Probably the most general attitude to take toward communication

[1] In *Jour. Acous. Soc. America, 22,* 689 (1950).

[2] Colin Cherry, *On Human Communication,* (New York: Technology Press of M. I. T. and John Wiley and Sons, Inc., 1957) , pp. 6 ff. This is a remarkably suggestive and comprehensive book and can be highly recommended to all who are interested in the subject. The present discussion owes much to it.

is to look upon it as composed of a series of signals or signs which can take on a very great variety of forms: they can be *audible,* as in speech communication, or in signals like drum beats, pistol shots or the strange sounds that often emanate from a radio loudspeaker; they can be *visual,* as in written language or in pictures (stationary or moving); they can be *tactual,* as in a kiss, a handclasp, or a punch on the nose. As we reflect on the matter, it is remarkable to note the extent to which all waking human individuals in modern society are being communicated with by other individuals as they carry out life's activities in business, education, social affairs or just plain domesticity. Such an important characteristic of civilization demands scientific investigation and is itself subject to the influence of science in accelerating measure. We shall now proceed to study the subject from this angle.

Speech and Hearing. Speech Communication

The ability of the human being to produce sounds, beginning with a baby's first cries, is the basis of man's earliest sophisticated means of communication. In his book *The Expression of the Emotions in Man and Animals* (1873), Charles Darwin comments on how efficient the vocal organs in all kinds of animals, including man, are as a means of expression. Great emotion such as fear or rage or pain can lead to violent action of the muscles of the body, and this in turn produces explosive puffs of air from the lungs which, modulated by the vocal cords, set the air in front of the mouth into vibration, with what we call sound as the result. By a process of evolution whose extent can scarcely be measured, this at first more or less involuntary reaction to unpleasant stimuli became, so to speak, tamed or controlled. It seems clear that at an early stage of evolutionary development birds learned to sing. Darwin attributes this to the desire of the male to communicate with the female during the breeding season. It must have been early in man's history that he discovered the possibility of communication with other human beings by the controlled use of his voice, thus leading to speech, one of his more precious possessions, unique in any sophisticated sense to the human race.

We can understand speech and hearing only by paying some attention to the branch of science known as *acoustics,* a word which comes from the Greek, meaning "hearing." It has to do with the physical phenomena associated with sound. When you open your mouth and utter speech, experiment shows that you disturb the air immediately in front of the mouth by squeezing it into smaller volume and increasing the local pressure slightly above the normal atmospheric value, so slightly indeed that the change in ordinary conversational speech amounts to only about a millionth part of normal atmospheric pressure (i.e., only about 1 dyne/cm^2 out of 10^6 dynes/cm^2). The interesting thing about this "squeeze" is that it refuses to stay localized in front of the mouth. Air is an elastic medium and after being squeezed tends to expand. Hence, the portion of the air in front of the compressed part in turn gets squeezed, and this in turn passes on the effect to the next adjoining region. The result is that the disturbance spreads in all directions through the air and constitutes what the scientist calls a *wave*. If the disturbance ultimately reaches a human ear, it communicates motion to the ear drum, whence it is transmitted through the little bones to the nerve fibers in the inner ear and by some process still rather mysterious we say the perceiving subject hears the sound.

Sound then, physically speaking, is an elastic wave in a material medium. Like all waves it has a speed or velocity which depends on the medium and the temperature. In air at room temperature, the speed of sound is about 340 meters/sec. It is common experience that sounds may be loud or soft. We have already commented on the small change in pressure produced by ordinary conversational speech. A very loud sound might produce a change of pressure of 100 dynes/cm^2 or 1 part in ten thousand of standard atmospheric pressure. Actually, the intensity of sound, as it is technically called, is measured by the flow of energy in the wave per unit area (cm^2) normal to the direction of propagation. If we use the erg (or dyne cm) as the fundamental unit of energy, intensity is then measured in ergs/cm^2 sec. The expression for the intensity of a plane sound wave (one proceeding as a beam in one direction only) is

$$I = p^2_{max}/2\rho_0 V = \overline{p^2}/\rho_0 V \qquad (1)$$

where p_{max} is the maximum excess pressure, ρ_0 is the average density of the fluid medium (e.g., the air) and V is the velocity of sound in the medium; $\overline{p^2}$ is the average of the square of the pressure, sometimes called the mean-square pressure. Its square root is referred to as the root-mean-square (rms) pressure and is a commonly used quantity in acoustics. In place of the erg/cm² sec as the unit of intensity, it is customary to use the watt/cm², where the watt as the unit of power is defined as the joule/sec or 10^7 ergs/sec. Actually, in acoustics it is more common to use a relative scale of intensity. Thus, if two sounds have intensities I_1 and I_2, respectively, they are said to differ in intensity level by D decibels (db), where

$$D = 10 \log_{10} I_2/I_1 \qquad (2)$$

It is customary to use for I_1 the standard minimum audible intensity of about 10^{-16} watts (cm²), making a total range of 140 db from the threshold of minimum audibility to that of pain. These figures depend on frequency, as will presently be explained.

The difference between what is commonly called a musical tone and just ordinary noise is appreciated by everyone. It was learned a long time ago (as long ago as the time of Galileo certainly) that the sound of a musical tone has a definite regularity about it, characterized by the fact that the air is squeezed periodically at a definite rate called the frequency, which is the number of times per second (called usually cycles/sec) the excess pressure goes through a complete cycle of values from minimum to maximum and back to minimum. The higher the frequency, the higher the pitch of the tone.

Such tones with definite frequency are called harmonic and are the simplest sounds in human experience. The time for a complete cycle of pressure values to be completed is called the period of the harmonic wave. It is of course the reciprocal of the frequency.

Another important characteristic of a harmonic wave is the wave length, usually denoted by the Greek letter λ (lambda). It is the distance between two successive points in a spreading wave in which the disturbance is precisely the same and doing the same thing (i.e., getting either larger or smaller). The disturbance in a harmonic wave travels a distance of one wave length in one period, and the product of wave length and the frequency equals the wave velocity. If we denote frequency by f and wave velocity by V, this relation (probably the most important one connected with harmonic waves) becomes

$$\lambda f = V \qquad (3)$$

This says that for a sound wave in a given medium with fixed velocity, high frequency is associated with short wave length and vice versa.

The normal ear is not sensitive to sounds of all frequencies. In fact, the range of hearing lies between about 20 cycles/sec and approximately 20,000 cycles/sec. Sounds of lower frequency than 20 cycles/sec are called infrasonic, and those of frequency above about 20,000 cycles/sec are known as ultrasonic. Ironically enough, there is rather more interest in modern acoustics in ultrasonics, that is, the sounds no human being can hear, than in the so-called sonic range. But one must not jump to the conclusion that ultrasonic waves are not useful in communication because they are inaudible. They have wide use, for example, in underwater-sound methods of detecting submarines.

But here we are mainly interested in sounds which are heard by the ear and hence serve directly as a means of human communication. The auditory threshold, that is, the minimum intensity at which sound can be heard, depends for a given individual on the frequency. When averaged over a great many individuals who appear to have what is termed normal hearing, the dependence can be most simply expressed in the form of the graph in Fig. 6.1, commonly known as the Fletcher diagram. It is interesting to note that the sense of hearing is on the average most acute at a frequency of about 3000 cycles/sec. The sensitivity falls off on either side until below 20 cycles and above 20,000 cycles; there is no sensation of hearing, only feeling which approximates pain.

We have here been referring to the actual physical intensity of sound as defined in eq. (1) and measured in db, as in eq. (2). Obviously, the perceived *loudness* of the sound has some relation to the intensity, but investigation shows that it is not as simple as might be thought. Intensity is objective and can be measured by physical equipment not dependent on hearing, whereas loudness is subjective and must be defined in terms of some kind of "average" listener. Loudness, then, is a psychophysical concept and its measurement is a cooperative affair between physicists and psychologists. One common method of measurement uses a reference tone of fixed frequency but variable intensity. It is then raised to such a level that it appears to the listener to be as loud as the sound under test. The number of db through which the reference tone has to be raised above the

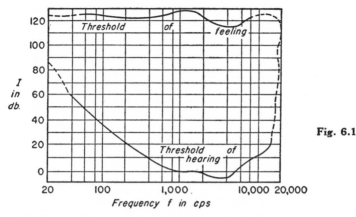

Fig. 6.1

From *Speech and Hearing in Communications,* by Harvey Fletcher. D. Van Nostrand Co., 1953.

minimum audible threshold of 2×10^{-4} dynes/cm² rms pressure in order to match the sound under test is called the *loudness level in phons.* The frequency of the reference tone is usually chosen as 1000 cycles/sec and the sound source placed at a distance of 1 meter from the head of the listener.

True loudness is still another thing. A scale for this has to have the property that when loudness of a certain number of units on this scale is multiplied by any arbitrary factor, the magnitude of the new auditory sensation then produced will be heard as equivalent to the corresponding number of units times that of the original sensation. The unit of loudness so defined is called the *sone.* It is defined to be the loudness produced by a tone at 40 db above the standard reference level of 2×10^{-4} dynes/cm². S. S. Stevens has established the following relation between loudness in sones (L) and intensity (I) relative to the minimum audible threshold (I_0) for the average listener.

$$L = 0.06 \left(\frac{I}{I_o}\right)^{0.3} \tag{4}$$

We can express this psychophysical law in terms of the loudness level in phons P (as defined just above), since

$$P = 10 \log_{10} \left(\frac{I}{I_o}\right) \tag{5}$$

Hence, if we take logs of both sides of equation (4), the result is

$$\log_{10}L = 0.03 \ P - 1.2 \qquad (6)$$

The logarithm to the base 10 of the loudness in sones is thus a linear function of the loudness level in phons.

It is worthwhile noting the physical significance of eq. (6). If the intensity I is 20 times I_0, so that $P = 13$ in phons, we have the loudness itself equal to 0.155 sone, while if $I = 40$ times I_0, with $P = 16$ phons, $L = 0.19$ sone. Thus, a doubling of the intensity leads to a loudness change of magnitude only some 0.04 sone. If $I = 80$ times I_0, L comes out to be 0.24 sone ($P = 19$ phons). To summarize, as the physical intensity I increases geometrically, the loudness level and loudness go up more or less arithmetically. This appears to be rather typical of the relation between stimulus and response in psychological phenomena. It is as though the receptor mechanism has a built-in device which prevents it from responding in direct proportion to the intensity of the stimulus. This may have developed in the evolution of man as a safety factor to prevent damage to the human sensory nervous equipment in the face of possible excessive energy flow in the environment. At any rate, it functions to a certain extent in this fashion, enabling the nervous system to cope with a much greater range of stimulus intensity than if the stimulus-response relation had been strictly proportional.

We must now turn our attention to speech sounds. These are formed when air is forced from the trachea through the vocal cords, which vibrate during its passage. This sets into vibration the air in the throat and mouth cavities and the air in front of the open mouth is thus periodically disturbed. It is a complicated business in which the tongue and teeth all play a role. The resulting compressional wave emitted is never a pure tone (save perhaps in the case of singing by an expert singer); rather, it is a complicated superposition of many different frequencies, giving it its recognized *quality*.

The intensity of speech naturally depends on the distance from the mouth. In ordinary conversational speech, the overall intensity can vary from about 80 db at 15 cm in front of the mouth to 65 db at 100 cm (these levels are of course with respect to the threshold of 2×10^{-4} dynes/cm^2). The power in speech sounds varies with the nature of the sound, being larger for vowels than for consonants, as common experience confirms. The vowels give volume to speech. On the other hand, the articulation or identifiability of speech sounds is due in large measure to the consonants.

It would take us too far afield to enter upon a detailed analysis of speech sounds, the study of which now forms a large part of modern applied acoustics.[3] Our interest in speech from the standpoint of communication resides essentially in its ability to convey information. Hence, it must be sufficiently intense and articulate. That is, the energy supply must be adequate to convey the wave disturbance from speaker to listener and the wave form of each word or part of speech must be sufficiently definite so that the listener identifies it correctly, i.e., as the speaker intended. Articulation therefore is of supreme importance and has received much attention. It is clearly a function of frequency; the essentially low frequency vowels do not provide much articulation, which is inherent, rather, in the higher frequency consonants. But articulation is also affected by the environment, notably by reverberation in a room and by noise. The ability of the ear to detect identifiable signals in the midst of noise, or generally confused and unwanted sound, is a great tribute to its sensitivity and selectivity. In this connection, the fact that human beings have two ears, or binaural hearing, may be noted. It has long been known that this enables us to a certain degree to locate the direction from which sounds come, certainly a very important element in auditory communication. But as Colin Cherry points out,[4] this result of binaural listening is really secondary to the ability conferred by the binaural effect to distinguish different sounds, e.g., a meaningful speech sound from a noise. The difference between the two is somehow enhanced (we are as yet by no means sure just how) by the difference in arrival at the two ears, e.g., difference in intensity, time, phase, etc., so that the ear and associated nerve stimulation can make the necessary distinction. It can be seen that the directionality capacity would follow from this more fundamental ability for image separation. Cherry in his article mentions numer-

---

[3] A good account may be found in Colin Cherry, *ibid.*, pp. 147 ff. See also E. G. Richardson, Ed., *Technical Aspects of Sound*, vol. 1 (Amsterdam: Elsevier Publishing Co., 1953), Chapter 10.

[4] Colin Cherry, "Electroacoustics for Human Listeners," *Journal of British Institution of Radio Engineers, 21*, 5 (January, 1961).

ous experiments shedding light on this matter. The practical importance of this can be considerable in stereophonic radio broadcasting and phonograph reproduction.

Communication and Control. Cybernetics

The obvious thing at this stage would be to embark on a study of the development of language as an illustration of speech and auditory communication. Language however is really a very complicated business, and hence we shall find it to our advantage to postpone its study in relation to science until we have paid attention to some rather recent ways of looking at the communication process in general. These have come about largely from two directions. The first is the purely technical problem of devising transmission lines which will transmit with high fidelity signals introduced into them at one end to a receiving device at the other. This has long been a vital concern of the communications engineer, and in endeavoring to solve it he has come up with a lot of fundamental ideas about communications in general. The second line of approach is of somewhat more recent origin, at least in its more sophisticated aspects. It is based essentially on the notion that the control of the behavior of an organism or mechanism involves communication. This is the discipline that has currently come to be called cybernetics.[5]

In any transformation of energy from one form to another, the control of the rate of transformation is a very important element in the process. When, for example, a group of workmen are laboring to construct a building, someone has to assume responsibility for controlling the rate at which the work is done. When engines replaced human labor as energy transformers, it became important to control the rate at which the engine would do mechanical work. Thus, when the load on the engine calls for larger power output, a control mechanism must assure that the engine knows this and increases appropriately the rate of transformation of energy from the burning fuel to the mechanical form. Now it does not take very profound analysis to see that in both cases the control takes place by someone or something telling someone or something else what to do. Communication is involved. In the first example, the laborer

[5] N. Wiener, *Cybernetics* (New York: John Wiley and Sons, 1948).

must be told, usually verbally by a supervisor, to work faster or slower. In the second illustration, the engine must be told what to do. In the early days of steam this was done by an operator manipulating levers so as to control the rate of flow of steam into the cylinder, and, for that matter, the manually operated throttle is still used on engines designed for the propulsion of vehicles at varying rates of speed, whether they be steam or internal-combustion engines. When you press your foot down or release it from the accelerator on your car you are communicating vital information to the engine in the endeavor to control the speed of the car, in so far as this in turn depends on the engine power output. (Of course, you also have brakes!)

To keep an engine running at steady speed independently of the load demands a degree of control which is irksome for a human operator. It is not therefore surprising that efforts were made in the early days of the steam engine to introduce automatic control. James Watt, who had much to do with the improvement of the early steam engine, invented for this purpose a clever device which he called a governor. This consists of two heavy metal balls attached to rods in such a way that they can swing about a vertical axis as a conical pendulum. The rotation is produced by attachment through gears to the driving shaft of the engine. When the engine speeds up, the balls whirl around faster. But since they are free to swing toward or away from the axis, the increased velocity of rotation is associated with increased centrifugal force which drives the balls further away from the axis, and by a connection with the throttle cuts down the rate of flow of steam to the cylinder and so tends to slow down the engine. The balls then drop slightly and the engine speeds up again. If the governor is properly adjusted, very delicate control of the speed is assured as only a very slight increase in speed will activate the mechanism. The details are not important for our purpose. The important thing is the nature of the role of communication in the control process. In this case, it has the striking characteristic that a small amount of the energy output of the engine is used to communicate to the engine the information that it ought to slow down or speed up, as the case may be. This is a process known as *feedback,* in which in the general case an energy-transforming or -transmitting system is self-controlled by the feeding back into the system of a small part of its output. A simple example of a control operating on the feedback principle is provided by the ordinary domestic thermostat. A small amount of the thermal output of the heating system is used to actuate what amounts to a switch. The latter closes or opens an electric circuit which in turn controls the burning of fuel in the furnace. Once again we see that the communication of information enters vitally into the picture; the room, so to speak, tells the thermostat that it is too warm; the thermostat conveys this information to the heating system by closing a switch and appropriate action results. In such cases, what may be called a small amount of information, or rather information communicated with a very small amount of energy, can serve to control the output of a very much larger amount of energy. Biological illustrations abound in the homeostatic mechanisms in the human body, whereby temperature control is assured over a certain range of environmental temperatures.

There is indeed another point of view about the connection between a control mechanism like a thermostat and the communication of information. If, for example, a heating system is running with the burner on and the room temperature is rising, that very fact informs the occupant of the room that the system is functioning (even if he does not hear it or otherwise sense it). But if the thermostat functions properly and by controlling the heating system maintains the temperature constant, the occupant remains ignorant of the action of the system. He is in a state of complete temperature equilibrium, at any rate, through any sensation of temperature—such information could come only if the temperature were to undergo appreciable fluctuation.[6]

It is clear that the experiences involving communication and the associated concept of control occupy a very large domain. The assignment of the name cybernetics (from the Greek *kybernetes*, a steersman) is therefore wholly appropriate. Actually, this name was apparently first introduced in the form "cybernetique" in 1834 by the French physicist André Marie Ampère in his *Essai sur la philosophie des sciences*. He used it to define the science of government or control, i.e., gave it a political meaning. Wiener, in reviving the

[6] This point is well brought out in W. Ross Ashby, *An Introduction to Cybernetics* (London: Chapman & Hall, 1956), Chapter 10.

term, has of course enormously extended its field of significance. Its biological applications extend to the study of the entire nervous system, including the brain. In the meantime, the development of large-scale computers for the rapid performance of mathematical calculations has stimulated study of the relation between such devices and the human brain. Automatic control through so-called "servomechanisms" is leading to a wholly new concept in industrial operations and we are due to hear more and more about "automation" and its consequences for society. The intensive study of all such things is a part of the province of cybernetics.

But here we are interested primarily in that branch of cybernetics which deals with the communication of information. This has now come to be called "information theory" and will be briefly surveyed in the following section.

Information Theory

To discuss the contemporary theory of information adequately, we must first be clear what "information" means in the context of the theory. To the layman, the word obviously conveys the significance of a message meaningful to a human being. But the word "meaningful" is fraught with so much vagueness that the people who decided to devise a scientific theory of information deliberately refrained from injecting this aspect into the subject. They were content to look upon information from the standpoint of the transmission of identifiable signs. Reduced to its lowest terms, information then becomes the selection of signs from a given list. To illustrate, suppose a given transmission system (e.g., a telephone circuit) can transmit the signs corresponding to the 26 letters of the English alphabet. The measure of information transmitted is plausibly connected with the ability to identify *one* of the letters out of the possible 26 that might have been sent. Thus, definite information is transmitted when the recipient unambiguously identifies the received signal, and the identification really consists in the ability to distinguish the sign from others that might be expected. This process is always a kind of guessing game and therefore probability is involved. It plays a considerable role in information theory.

To be more specific, let us assume that we have available for the transmission of information C characters, signs or symbols. These might, for example, be the 26 letters of an alphabet, the larger number of letters and other symbols (capitals, lower case letters, punctuation marks, etc.) found on the keyboard of a typewriter, the dot and dash of the Morse code, the numbers 0, 1, 2, – – – 9, or just the two symbols 0 and 1. Many other possibilities will occur to the reader. The principal point is that the number shall be finite and that each sign can be readily distinguished from any of the others. Suppose now that each message transmitted is made up of L signs. This might be a single word of L letters or it might be a collection like 0110011. How many different messages can be transmitted under such circumstances? In each message the first sign can be chosen in C different ways, since any one of the C available signs can be used first. But the second sign in the message can also be chosen in C ways, and so can the third, fourth, etc., up to the Lth. Therefore, the total number of different messages is the product of C by itself L times or

$$N = C^L \tag{1}$$

It has become customary to turn this around, so to speak, express L in terms of N and C, and say that the amount of information necessary to convey a message of length L using C different signs or symbols is

$$L = \log_C N \tag{2}$$

Put it another way: let us suppose we know there are N different possible messages, all of the same *a priori* probability. If C signs are available for the transmission of these messages, then the amount of information necessary to convey each message, which will be interpreted as the information per message, will be defined as

$$\log_C N \tag{3}$$

We can make this more meaningful by taking a very simple concrete example. Let us suppose that the transmitted message is one of the first 16 capital letters of the alphabet, A to P, and further let us assume that we are going to use just two symbols, + and −, to transmit the messages. This means, in terms of a previous notation, that $C = 2$ and the number of possible messages is $N = 16$. Hence, the amount of information per message is

$$\log_2 16 = 4$$

This means that each message, in order to convey the information that one of the 16 letters is meant, must contain 4 of the symbols + and —. As a matter of fact, we can see this very quickly by a mere process of counting: there are indeed 16 and only 16 different arrangements of + and — taken 4 at a time. The assignment of each arrangement to each letter is arbitrary, but a simple way is as follows. Of any letter we can ask ourselves: Is it in the first or the second half of the whole set? If the former is the case, we assign + for the first symbol of the message, otherwise —. The second symbol is assigned by asking whether the letter in question is in the first half of the first half. If so, again we assign +, otherwise —. If we pursue this little game, we get the assignment in Table 6.1.

TABLE 6.1

A	+	+	+	+
B	+	+	+	—
C	+	+	—	+
D	+	+	—	—
E	+	—	+	+
F	+	—	+	—
G	+	—	—	+
H	+	—	—	—
I	—	+	+	+
J	—	+	+	—
K	—	+	—	+
L	—	+	—	—
M	—	—	+	+
N	—	—	+	—
O	—	—	—	+
P	—	—	—	—

It is seen that the various different combinations of + and —, 4 at a time, are just sufficient to represent unambiguously the 16 letters.

Hence, it is proper to say that 4 is the amount of information necessary to transmit the message in question using the code composed of two symbols only. This is called a binary code. Its simplicity and adaptability to transmitting systems have conferred on it such significance that most modern practical information theory is based on it. It is indeed used to define a unit of information, namely, the *binary* digit or *bit*. Thus, each message in the example described has an information content of 4 bits, since it takes a combination of 4 of the + and — symbols to identify the message. In what follows, we shall assume the use of the binary code and hence define the information of each of N equally likely messages H as

$$H = \log_2 N \tag{4}$$

Note the phrase "equally likely." This is evidently an ideal use; we cannot expect that in the transmission of information in an ordinary language every possible message will actually be equally likely; some messages are certainly more frequently encountered than others. We recall that in the English language, for example, the letter E occurs more frequently than any other in the average passage of prose. Before we tackle this problem, however, there is another more obvious one staring us in the face: What happens when the number N is not a power of 2? In our example, we were careful to choose 16 letters as our possible messages, and since $16 = 2^4$, this gave the integral value of 4 for H, with the result that precisely 4 bits of information, i.e., a combination of 4 + and — signs, are required to transmit the messages. But suppose that $N = 19$, i.e., that we add 3 more letters to the list. It is clear that there is no integer n such that $19 = 2^n$. In fact

$$\log_2 19 = \log_{10} 19 / \log_{10} 2 = 4.25$$

Now there is obviously no meaning to four and a quarter symbols forming a message! There are, however, two possible interpretations, which still permit us to use this non-integral quantity as a measure of information. In the first place, we see that though 4 symbols will be inadequate for the representation of all possible messages, 5 will do it and will be actually more than enough. Hence, to say that the information is 4.25 bits does convey useful information in itself. The second possibility involves the introduction of time and the rate at which information is transmitted. After all, in any transmission system, whether it be human speech and hearing or an electrical transmission line, the rate is an important item. Suppose that in the case just discussed Q messages can be transmitted on the average per second. Then the amount of information passed along in bits per second is

$$H' = Q \log_2 N \tag{5}$$

Even H' may be a non-integral number. But it can still have meaning as an *average* rate of information transmission, and over a long enough time it will of course yield an integral number of bits. Thus, it might be that for N = 19, Q = 10 messages/second on the average so that

$$H' = 42.5 \text{ bits/second}$$

In 10 seconds, therefore, the total information becomes 425 bits, i.e., 425 binary choices are necessary to send in 10-seconds messages at the average rate of 10 per second, each of which consists of one of the 19 letters A to S. It must once more be emphasized that all messages are assumed to have equal probability.

It is now necessary to see what can be done to modify the above simple considerations when different letters or messages have different probability of occurrence. If the reader will examine a few pages of ordinary English prose (a newspaper, for instance), he will find that the average frequency of occurrence of different letters is fairly constant. Thus, the letter E occurs on the average about three times as often as the letter L. Similarly, A occurs on the average four times as frequently as W. Printers have long recognized this average frequency by providing type in amounts proportional thereto, and corresponding to the order E T A O I N S H R D L U C M F W Y P V B G K Q J X Z. Recognition of this order speeds up typesetting. From the standpoint of our present concern with information theory, it is clear we must modify the expression for the information involved in the transmission of messages consisting of letters in the alphabet to take account of the different average frequency of occurrence associated with them.

We shall use a method given by Cherry.[7] Let the *a priori* prob-

[7] Colin Cherry, *On Human Communication, op. cit.,* pp. 179 ff.

abilities associated with the various letters A, B, C – – – be denoted by p_a, p_b, p_c – – –. These are the relative *fractional* frequencies of occurrence of the various letters, i.e., if $p_a = \frac{1}{4}$, it means that the letter shows up on the average for every 4 messages. Since some letter must show up in every message, we have

$$p_a + p_b + p_c - - - + p_z = 1 \qquad (6)$$

Suppose we consider an aggregate of all the possible message sequences each of L letters in length. Let the number of such be n, and assume that the sequences are long (L $\gg$ 1). The relative fractional frequency of each sequence (or its probability, as we may call it) is

$$p(L) = (p_a{}^{Lp_a})\,(p_b{}^{Lp_b})\,(p_c{}^{Lp_c}) - - - (p_z{}^{Lp_z}) \qquad (7)$$

For the letter A occurs on the average Lp_a times in each long sequence, since p_a is the probability for its single occurrence. But we also know that if the probability for one occurrence of A is p_a, the probability for 2 occurrences in the same sequence is $p_a{}^2$, for three occurrences $p_a{}^3$, etc., so that the probability for Lp_a occurrences is $p_a{}^{Lp_a}$. We therefore get the joint probability for the whole sequence, i.e., the probability that A, B, C – – – occur, as the product given in eq. (7). From the meaning assigned to p(L) we have, if we can assume that all the sequences of n messages are very nearly equally likely (and the longer the sequences are, the more nearly plausible this becomes)

$$n = 1/p(L) \qquad (8)$$

Moreover, the information per sequence is now simply

$$\log_2 n$$

or

$$\frac{\log_2 n}{L}$$

per sign. Hence now

$$H = \frac{\log_2 1/p(L)}{L}$$

$$= - \frac{\log p(L)}{L}$$

$$= - \frac{\sum\limits_{i} \log p_i{}^{Lp_i}}{L} \qquad (9)$$

where the summation over i means summation of the various terms in a, b, c, − − − z. Since

$$\log p_i{}^{Sp_i} = Sp_i \log p_i$$

we have at once

$$H = -\Sigma p_i \log p_i = -(p_a \log p_a + p_b \log p_b + - - - + p_z \log p_z) \quad (10)$$

for the number of bits of information per letter or sign.

This formula looks a bit mysterious, especially in view of the negative sign (but reflect that the p's are all proper fractions and the log of a proper fraction is always negative). Perhaps it will not look so formidable if we examine what it becomes for the special case in which all the p's are equal, i.e., go back to the original simple problem. For this, $p_a = p_b = - - - = p_z = 1/N$, where N is the total number of signs. Then

$$H = -\frac{1}{N} \Sigma \log 1/N = \log N$$

as before. Note that all logarithms here are to the base 2. It is of interest to note that it is precisely for the case in which all the p's are equal that the information is a maximum for the given set of messages. In other words, it takes more bits per message in this case than for that in which some signs are more likely to occur. It does not take much mathematical knowledge to see the plausibility of this. It is of course closely connected with the construction of language as a means of communication in such a way as to minimize the amount of information needed to convey a given message. Recall in this connection the discussion of Zipf's law in the preceding chapter.

Our purpose in this section has not been to develop in detail the mathematical analysis of information theory, but rather to give the reader some impression of the fundamental concepts and methods used in the development of the subject, which already has a vast literature associated with it. To see that we have just scratched the surface, it is enough to recall that we have limited ourselves to discrete messages like single signs or letters of the alphabet. But most communication is actually done by means of continuous radiation. Thus, in speech we do not emit single clear-cut sounds in staccato fashion. We just talk and the various sounds form a succession of very complicated wave forms merging continuously into each other. It becomes an interesting problem to see how such a continuous stream of signals can be broken down into a set which can be handled by information theory. This requires more elaborate mathematics than we can go into here. The interested reader will find about as clear an account as any in Cherry's *On Human Communication,* referred to previously. Suffice it to say that the difficulty can be overcome. It has led to a detailed study of communication over electrical transmission lines and the study of the problem of the proper coding of messages for given types of communication channels, that is, the preparation of signals which can be handled by the channels available. Another problem vital for information theory is the influence of extraneous factors on the transmission of messages. These influences which act to distort or degrade the transmission are known collectively as "noise." In the case of airborne sound, the significance of noise as any unwanted sound which interferes with the intelligibility of the signal it is desired to transmit is immediate; we all know how difficult it is to hear clearly and understand a person with whom we are conversing at a cocktail party because of the general noise due to all the conversations going on. The term has been applied more generally to all disturbing influences in communication systems. Thus, random electron motions in a vacuum tube will introduce "noise" into an electrical amplifier of which the tube is a component part, just as static electricity in the atmosphere adds noise to the electromagnetic radiation transmitted to a radio or television set. In any communication system, the so-called signal-to-noise ratio therefore becomes a vital matter, if the signal is to be identified without error. Noise clearly increases the amount of information necessary to convey a message.

We shall come back to these matters when we examine the nature of ordinary language as a means of communication. But first there is a very important connection between information theory and the branch of physical science called thermodynamics which we must explore, as it promises to have far-reaching implications.

Information Theory and Thermodynamics. Entropy

That there should be any connection between the communication of messages as studied in information theory and the theory of thermodynamics may occasion considerable surprise. The latter discipline is ordinarily thought of solely in connection with the work of heat engineers and physical chemists and has the reputation of being an extremely difficult subject, particularly among young students of physics, chemistry and engineering who rarely understand it without having their noses rubbed in it several times. Actually, it is one of the most general and all-embracing scientific theories ever invented, and in its scope can cover all the transformations of matter both living and nonliving. It is of such great significance to every human being that some acquaintance with it is imperative for every thoughtful person. Fortunately, such an acquaintance can be arrived at without the use of elaborate mathematical analysis. Since we plan to make use of thermodynamic concepts and applications in much of the rest of this book and since it is relevant to the material of this chapter, we shall digress momentarily to develop some ideas about it here.

The suggestion of the possible connection between information theory and thermodynamics actually came out of the similarity in form between the expression H in eq. (10) and a thermodynamic relation involving entropy, one of the leading concepts of thermodynamics. However, it turns out that the relation can be made clear on the basis of another kind of analogy, and while analogies in science can often be tricky and deceptive, we shall find it worthwhile to pursue it here.

But first we must try to describe the nature of thermodynamics. As has been suggested, it is a theory of great generality since it can be applied to matter of all kinds without regard to its constitution. In the form which we discuss here, it relates to matter in so-called states of equilibrium. A state of equilibrium is one such that if a particular body, say some water in a tea kettle, is left by itself undisturbed by outside influences, it will not change. Suppose we leave the tea kettle on the stove full of water with the cover on but do not light the gas or turn on the current through the burner (if it is an electric stove). We go away and leave it for a while. If on our return we find the tea kettle still there and still full of water at the same temperature as before, we shall say it has stayed in a state of equilibrium. If, on the other hand, we light a fire under the kettle, the water gets hot and may boil away. In this process it is not in a state of equilibrium. But in this case it is subject to outside influences. Can objects change without some influence from the environment? Yes. Imagine a glass of water in the bottom of which a crystal of copper sulphate is placed. The system starts out as a mass of colorless water and a bit of blue crystal. If we carefully shield it from the surroundings and go away for a while, when we return we shall see the blue color spread throughout a large part if not the whole of the glass. We say the crystal has dissolved in the water. Here is a system which is not in equilibrium. A little thought is enough to persuade one that real states of equilibrium are actually difficult to realize in our imperfect world. Nevertheless, we can approximate them and that justifies us in inventing the concept even if it is an ideal. This is the sort of stuff of which scientific theories are made, as we have already emphasized in Chapter 1.

So we shall say that thermodynamics is the science which deals with matter in states of equilibrium and with the changes from one equilibrium state to another. Such a change is called a thermodynamic process. The significance of the name arises from the fact that in many thermodynamic processes the temperature changes, i.e., heat is somehow involved, and at the same time mechanical work is done either on or by the system, as for example, in the exploding gas in the cylinder of an internal-combustion engine. However, we must not think thereby that thermodynamics deals exclusively with heat, for there are thermodynamic processes in which no change of temperature takes place and there are others in which no work is performed. The name therefore fails to do justice to the generality of the discipline.

Thermodynamics operates with two great concepts and in terms of two very general hypothetical principles. The concepts are those of *energy* and *entropy,* and the principles are the so-called first and second laws of thermodynamics—better called principles, since they are not scientific laws in the sense of Chapter 1, i.e., they are not directly observed regularities or patterns of experience. It is true that they have been deduced mathematically from a more general theory called statistical mechanics, and in this context function as laws in the sense of our earlier discussion. But into this we refrain from entering at the moment.

The concept of energy is the embodiment of the attempt to find in the world of our experience an invariant, or something that stays constant in the midst of the flux of change. The human mind from the time of Parmenides (a Greek philosopher of the fifth century B.C.) on down has felt an irresistible urge to find a sheet anchor that would, so to speak, stay put while everything was apparently changing. The concept first grew up in connection with the motion of bodies, e.g., with the search for something constant about the motion of a simple pendulum which swings ceaselessly back and forth. Offhand it would seem as if the search would be fruitless, until one observes that as the bob falls it gains speed, and as it rises again, it loses speed. This suggests the description of the motion in terms of two quantities, the first depending on the speed and the second on the position of the bob relative to its lowest position. We can find two such quantities such that though they themselves change continuously, their sum is always *constant* in time. This is called the *mechanical energy* of the system constituting the pendulum. Unfortunately for the science of mechanics, simple observation shows that it really does not stay constant, for the bob always ultimately comes to rest if left to itself. However, this has proved fortunate for science as a whole, for it challenged men to look for a reason why, and they found it in the appearance of other physical effects, e.g., heat, as the mechanical energy apparently disappeared. This suggested an enlargement of the concept until it now covers the whole gamut of physical experience. We think now of energy appearing in many forms besides the mechanical form just mentioned, such as heat, chemical action, light, sound, electric current, etc. Whenever an effect corresponding to any phenomenon of these and other similar kinds (encompassing indeed all the phenomena in the observed universe) occurs with the appearance of a certain amount of energy in the corresponding form, this is always accompanied by an equivalent loss in energy in some other form, so that the total remains constant. This is the content of the first law of thermodynamics, the general principle of the conservation of energy. In simple language, it expresses our conviction, based on a mountain of experience, that in no part of life's experience is there any such thing as getting something for nothing. No matter how sophisticated life becomes in the complexity of energy transformations in which we are immersed, we appear to be faced with this inexorable situation: In no way can we create energy; we can merely transform it, from the fossil fuel or uranium in the ground to the heat and electrical energy which warms and lights our homes and keeps us all forever on the move, "forever" here meaning of course as long as we can find the wherewithal to transform. It has now become a truism that energy is the basis of civilization on earth: in the metabolism in our bodies which sustains us as so-called living creatures, in the activity of our brains which enables us to reflect on our doings and even invent the concept of energy as a convenient way to describe them, and in all human experiences without exception. Small wonder that energy is the primary concern not only of scientists but of all who are involved in the study of any form of human activity, even though in many cases they do not realize it and use another terminology.

But the first law of thermodynamics is not the whole story. Experience indicates that in the transformations of energy which keep life and civilization going, a certain penalty is exacted. This somewhat melancholy sequel to the first part is the content of the second law of thermodynamics, which says that in every naturally occurring transformation of energy there takes place somewhere a loss in the availability of energy for the future performance of work. In other words, every transformation reduces by so much the practical possibility of future transformations, even though they would be theoretically possible so far as the first law is concerned. Every natural transformation indeed involves a kind of one-wayness which somehow seems definitely associated with the notion of things running down. Consider for example two pieces of metal, one of which is at a higher temperature than the other. If we bring them together, what happens? So far as the first law is concerned, it is just as reasonable that heat should flow from the cooler body to the warmer body, at any rate until the cooler body had lost all its heat, as it is that heat should flow from the hotter to the cooler: there is no change in the total energy in either case. Actually, experience shows that the heat tends to flow always from the hotter to the cooler body and indeed stops flowing only when they have reached the same temperature. That there results thereby a loss in the availability of energy is made manifest by the fact that whenever a temperature difference exists between two objects, it can be made to produce useful mechanical work as in a heat engine. Every heat engine functions ultimately by the passage of the working substance (e.g., steam, in the steam engine) from a high temperature (e.g., the boiler) to a lower temperature (e.g., the exhaust). When such a temperature difference disappears by a natural flow process, the chance to obtain mechanical work from it also disappears.

The loss of the availability of energy is also evident in the operation of a heat engine itself. It is well known that such an engine transforms energy in the form of heat (produced by the combustion of fuel) to energy in the form of mechanical work, e.g., the motion of wheels. So far as the first law of thermodynamics is concerned, it should be possible to change *all* the heat in a given mass of working substance into work. But no actual engine does this. In each cycle of the engine it is apparently necessary to sacrifice a certain amount of heat, which is discharged through the exhaust at a lower temperature than that at which it is taken in. We express this by saying that no heat engine ever functions continuously at 100 percent efficiency.

The German physicist Rudolf Clausius introduced the concept *entropy* to provide a quantitative description of the loss in availability of energy in all naturally occurring transformations. In terms of this concept, the second law of thermodynamics states that, for a closed system, that is, a physical system cut off from contact with the surrounding environment, the entropy cannot decrease: it must increase or stay the same. It should be emphasized that though increase in entropy measures the loss in availability of energy, it is itself not a quantity with the physical dimensions of energy; rather, its dimensions are equivalent to energy divided by temperature. But we need not pursue this precise definition in order to understand the significance of entropy in energy transformations.

Clausius was not content to limit the application of the principle of entropy, as the second law can be phrased, to closed systems. He felt it ought to be generalized to the whole universe, by assuming that in any naturally occurring process the *tendency* is for the entropy of the universe to *increase*. Note, of course, that in local environments we can make entropy *decrease* by providing external compensation. Thus, though the natural tendency is for heat to flow from a hot to a colder body with which it is placed in contact, corresponding to an increase in entropy, it is perfectly possible to make heat flow from the cold to the hot body, as is done every day in a mechanical refrigerator. But everyone knows that this costs money! It does not take place naturally or without some extra effort exerted somewhere (e.g., in the electrical power station). The most careful examination of all naturally occurring processes (that is, those in which artificial external influences are not allowed to intervene) has only served to confirm our assurance in the inexorable overall increase in the entropy of the universe. Every time anyone lights a cigarette, the availability of energy in the universe goes down and the entropy goes up. Clausius very neatly epitomized the fundamental principles of thermodynamics: the energy of the world stays constant; the entropy of the world increases as if striving to attain a maximum value. If the essence of the first law in everyday life is that we cannot get something for nothing, the second law emphasizes that every time we do get something we reduce by a measurable amount the opportunity to get that something in the future, until ultimately the time will come when there will be no more "getting." This is the so-called "heat death" envisioned by Clausius, when the universe will have reached a dead level of temperature throughout, and though the total amount of energy will not have changed one jot, there will be no means of making it available, that is, of producing a transformation which will raise the temperature of one part above that of the surroundings—the entropy will have attained its limiting maximum value. To some a melancholy outlook! Of course, it hardly pays to get unduly excited over it, since the time scale for the degradation of energy in our universe is enormous, and it is difficult for most of us to worry much over the fate of our great-great grandchildren, much less exhibit concern over what is likely to happen to our world a billion or so years from now. Still, it *is* something to ponder!

The extension of the principles of thermodynamics to the universe is a somewhat drastic extrapolation of what is after all a postulate which appears to be successful through its deductions in our own immediate vicinity. Some may question the desirability of this extrapolation and may view the "heat death" as a wholly unnecessary nightmare. Yet there is a point of view with respect to the foundations of thermodynamics which makes it all too plausible. This point of view is the one called statistical, in which one gains knowledge about complex systems containing large numbers of entities largely by virtue of the greatness of the number and without specific acquaintance with the properties of the individual members. We have already called attention to the life-insurance tables which are able to predict about how many persons in a given population and of a given age range will die in a certain period of time, without being at all able to say which ones. Or, to take a physical example, consider the molecules in a cubic centimeter of gas confined at room temperature and normal atmospheric pressure. Their number is of the order of 10 billion billion. We cannot hope to tell much of anything specific about each one, but just because there are so many, we can predict roughly how many will be moving on the average at any particular speed. This may seem rather mysterious until we reflect that we can always think of any aggregate of entities as changing (if it changes at all) in such a way as to make its subsequent condition more likely than its earlier one. Changes in nature proceed from less probable to more probable states.

If we had some way of defining the probability of the state of an aggregate it would therefore appear plausible to define the entropy of the aggregate as some function of the probability, so that as the probability increased the entropy would also increase. Then in all naturally occurring changes the entropy would most likely increase, and this would provide a logical basis for the second principle of thermodynamics. Note, of course, that it does not imply that the entropy as so defined has necessarily to increase in any natural change, it simply is *likely* to do so and the probability of this likelihood can also be computed.

Suppose, to take a simple example, that all the air in a hermetically-sealed room could be gathered together in a small bottle in one corner and the stopper inserted. This would take some doing, but there are appropriate devices for this purpose if we are willing to expend the necessary energy. Now suppose the stopper is removed. No one with the simplest experience of natural events questions that the air in the bottle will rush out and fill up the room again. Numerous explanations can be given depending on the sophistication of the observer (making use of the physical ideas of force and pressure and the like), but the simplest of all is probably that which says that if the molecules of the air have a choice of staying cooped up in the small bottle or roaming around the much larger room there is greater likelihood that they will prefer the latter. If this seems too anthropomorphic, we can say that there are many more ways in which the molecules of the air can distribute themselves among the various elements of volume of the room (say all equal to the volume of the bottle) than there are those in which they can stay in the bottle. Thus, of any one molecule we can say that there are many ways in which it can locate itself in a volume element of the room equal to the volume of the bottle contrasted to the one way it can remain in the bottle.

With this example, we have already decided on a measure of probability, namely, the number of different ways in which a given situation can occur. It is a particularly simple and straightforward one since it depends ultimately on the ability to count. The more ways a given state can be realized the greater is its probability (sometimes referred to as the statistical probability, to distinguish it from the probability of a given event which is usually represented by a proper fraction, as when we say that the probability of throwing a 2 when a die is cast is 1/6).

The choice of a constant times the logarithm of the statistical probability of an aggregate as the measure of the entropy of the aggregate was the great contribution of Ludwig Boltzmann to thermodynamics, for it leads directly to the second principle on a statistical basis and provides the statistical foundation for this whole discipline. From this definition we can indeed derive in logical fashion both the first and second laws of thermodynamics, and they then become scientific laws in the sense of our earlier discussion in Chapter 2.

We may epitomize Boltzmann's definition in the simple formula

$$S = k \log W + c$$

where S denotes the entropy, W is the statistical probability, k is a universal constant of nature, called, appropriately, the Boltzmann constant and having the value 1.37×10^{-16} ergs/^{o}C. The quantity c is an arbitrary constant, whose value depends on the particular circumstance of the problem under discussion. We need not worry about it here. The essence of this formula was placed on Boltzmann's tombstone.

The statistical interpretation of entropy and the principles of thermodynamics leads at once to another important point of view, namely, that of the role of *order* in the direction of naturally occurring processes. It does not take much reflection to see a profound connection between statistical probability and order. Let us take, for example, an aggregate of atoms in a closed rectangular space of volume V and ask ourselves the number of ways in which these atoms can be distributed uniformly in the space, i.e., so that n equal-volume elements of the space, irrespective of position, shall contain on the average the same number of atoms. If there are N atoms (where N $>>$ 1) the approximate expression for the number of ways becomes

$$W = (2\pi N)^{\frac{1}{2}} n^N (n/N)^{n/2}$$

For example, if $N = 10^{20}$ and $n = 10^{10}$, this number is approximately 4.6×10^{21}. This may justly be termed a random arrangement. Suppose, on the other hand, we desire to distribute the atoms so that there is one of them at each point of a regular lattice formed by three sets of mutually perpendicular planes parallel, respectively, to the faces of the enclosure. The lattice consists of the points of intersection of these planes. The atoms are in this case distributed in a regular array simulating a crystal. This is a much more orderly distribution than the one first contemplated and common sense suggests that for a given number of atoms, given volume and given number of lattice points (of order of n above) there are far fewer ways of realizing it. Precise calculation confirms this conjecture, indicating that in this problem, at any rate, the greater the number of ways of realizing a given distribution, the smaller is the degree of order associated with it, and conversely. This turns out to be the case in general. We can therefore introduce another interpretation of entropy. Increase in entropy means a transition from a more orderly state to a less orderly state.

Let us take a more familiar illustration, namely, the distribution of playing cards in the normal pack. Everyone knows that with well-shuffled cards the result is in general a random distribution, with small chance of finding the cards in a given suit arranged in the order considered regular by arbitrary choice of the people who use the cards for playing games. The latter more orderly arrangement is simply statistically less probable. It turns up far less often; there are relatively fewer ways of realizing it. Once again, high statistical probability is associated with relatively high randomness or disorder, while small statistical probability is correlated with relatively great order.

On this interpretation, the meaning of the second law of thermodynamics, the law of increasing entropy, is now clear. In any naturally occurring process the tendency is for all systems to proceed from order to disorder, and the maximum entropy of Clausius is the state of complete disorder or thorough randomness, out of which no return to order is practically possible because it applies to the universe as a whole and nothing short of an inexpressibly improbable revolution could reverse the process and decrease the entropy. From this point of view, the trend from order to disorder with production of entropy is inexorable. The second law always wins in the end. A gloomy outlook indeed! But there is perhaps a silver lining in the cloud.

We have already commented that local decreases in entropy are possible, provided one is willing to pay the price. The most striking is that manifested by the living organism. The production of a living creature on no matter how humble a level is a vivid example of the transformation of disorder into order. From a random collection of atoms of oxygen, nitrogen, hydrogen and carbon, with a few others thrown in, are synthesized the remarkably elaborate chemical constituents of the living cell, and these cells in turn arrange themselves functionally in intricate but orderly fashion. Here we must avoid confusing complexity with disorder. The living cell is enormously complex in terms of the constituent molecules, but these molecules are models of orderly arrangement. Life then may fairly be said to *consume* entropy, since with the transition from disorder to order, the entropy of the universe decreases. Some entropy disappears: we shall refer to this as a process of consumption. It corresponds equally well to the production of a highly improbable configuration from a more probable one. Certainly the creation of a living organism is such a move in the direction of the improbable. Nothing but living things themselves have so far been able to negotiate this synthesis. The production of life is of course reproduction.

We must be careful to point out that the consumption of entropy in the formation of living things does not constitute a violation of the second law. So far as we are aware, this consumption is accompanied by the corresponding production of entropy elsewhere in the universe; at any rate, it seems altogether likely. But whether this is so or not is not the important point here; the significant thing is that in the living organism there is an entropy consumption by very virtue of the transition from disorder to order. If we were mathematically clever enough, we could calculate it and doubtless someday will.

The entropy consumption associated with the reproduction of life may be thought of as unconscious or at the most instinctive. Rather more interesting is that connected with the conscious functioning of the human nervous system. The development of civilization itself may be in a certain sense looked upon as the result of the attempt of man to introduce order into his environment. On the physical side, he builds structures for shelter, transport, and for the construction of other structures, i.e., machines of all kinds. All these imply creation of order out of disorder: a house is more than a pile of bricks or stones; it is an *orderly* arrangement of component parts and thus represents a decrease in or consumption of entropy. On the social side, man creates language, an orderly pattern of speech sounds and written signs. The communication of information from man to man replaces with order what would otherwise be chaos in the relations of individuals. The social institutions of government, law, and education reflect the same entropy-consuming tendency. In all these ways man endeavors consciously to maximize the amount of order in his environment and to consume as much entropy as possible.

The reader who may have begun to wonder just what this elaborate digression on the fundamental concepts of thermodynamics has to do with information theory, if he is sufficiently perceptive, has probably caught a glimpse of the connection in the preceding paragraph. For the communication of messages may obviously be considered an illustration of the imposition of order in an otherwise random assortment of signals. Hence, we may validly look upon information as measuring a decrease in or consumption of entropy. On the other hand, the presence of noise in the environment in which communication is taking place works in the opposite direction: it produces entropy by tending to transform order to disorder. If the reader does not care to think of entropy as something which can be locally produced and consumed (i.e., decreased), he can take advantage of a new term introduced by L. Brillouin,[8] namely, negentropy, for the negative of entropy. Then the transformation from disorder to order is said to increase the negentropy of the universe, while the transition from order to disorder, of course, increases the entropy.

The analogy between the transmission of information and the consumption of entropy, first suggested by eq. (10) on page 151 and then made qualitatively plausible by the considerations of the present section, has been applied in the analytical development of information theory by the applied mathematicians and communications experts. It has indeed been pointed out (notably by Colin Cherry) that this is by no means a necessary step. One can get along without it, as one can with all analogies, since all "safe" analogies in science are really mathematical in character. It is true that many qualitative analogies can be misleading, and scientists have often inveighed against philosophers for their careless use of such. Nevertheless, there is much to be said for the heuristic value of the cautious use of analogy in the discussion of an idea with people who wish to get some grasp of what is going on, even if they never intend to use the idea professionally to make detailed computations. It is on this basis that we have thought it worthwhile to introduce the thermodynamic interpretation of information theory.

The whole business has perhaps a larger significance in a wider context. We have already noted that all the human experience which is the subject matter of science comes to us in the form of external stimuli which are messages from what we may choose to call a world external to ourselves. All our experience is thus a problem in communication and involves the transformation and transmission of energy. But from the second law, this is always accompanied somewhere (possibly not locally—but somewhere) by an increase in entropy. In other words, the entropy consumption which corresponds to the im-

[8] L. Brillouin, *Science and Information Theory* (New York: Academic Press, 1956). See in particular Chapter 9.

position of order or disorder implicit in the message is compensated somewhere by an equivalent or indeed a greater entropy production. This holds no matter what the nature of the stimulus is. But a particularly interesting case is that of communication via electromagnetic radiation or light. It is well known that all objects emit such radiation in the form called thermal at all temperatures above absolute zero, and that the rate of radiation increases as the absolute temperature increases (in fact varies directly as the fourth power of the absolute temperature). Now the interesting point is this: our ability to tell two objects apart by the radiation emitted by them depends essentially on the existence of different rates of radiation for the two and this in turn depends on their being at different temperatures. As soon as they have come to temperature equilibrium (have come to the same temperature), we can no longer distinguish them. (It is here assumed of course that we get no clues from radiation coming from other sources and reflected or scattered by the objects in question.) In other words, with the increase in entropy associated with the attainment of equilibrium there comes about a loss in information; the system is able to say less and less about itself through the emission of radiation. The problem, when examined in greater detail, is a bit more complicated than we have made it out to be. But the ultimate result is plain: increase in entropy means loss in information.

The Nature of Language

Communication obviously demands some sort of code in which the communication can take place. We have already mentioned the value of the binary code in the transmission of information. Any such code may be said to constitute a language, provided that it conveys something understandable to a suitably oriented person. Put in other words, a language is a collection of signs for the purpose of communication. These signs may be speech sounds or they may be written symbols; or they may even be, for example, arbitrary orientations of human fingers. Gestures without speech or writing can constitute language, albeit we should say of a somewhat primitive kind. The American aborigines have used successfully as language a code of smoke signals. All that is ultimately demanded of a language is that it contain certain symbols which can be differentiated from each other (i.e., recognized apart), can be transmitted from one place to another by some means, and not only can be identified by a recipient but are also intelligible to him in terms of some assignment of meaning to them.

One can approach the study of language from several points of view. The first obviously is the history of the development of language—a fascinating chapter in the history of the human race but entirely too long a story even to be reviewed here. From the standpoint of science, a more obvious line of attack is the logical structure of language, closely connected with general information theory. This study can of course lead to a consideration of how any particular language can be improved by artificial means or how entirely new languages can be created for certain purposes.

We are going to confine our attention in this brief review to the logical structure of language and its relation to culture. But there is one point which must be made clear at the outset. If we are to discuss language we must have a language in which to do it, and this can hardly be identical with the language being discussed. Hence arises the need for developing a metalanguage to describe the ordinary object—language of everyday life. This need has been met by the theory of signs, called by C. W. Morris[9] *semiotic* (the e is long, since the word has been coined from the Greek, in which it replaces *eta*).

The principal subdivisions of semiotic are syntactics, semantics and pragmatics. Syntactics is the study of signs and the relations between them. Semantics discusses the relations between signs and the things designated by them. Pragmatics is the study of the relations between signs and their users. Already we see the metalanguage creeping in! However, the terminology is not too forbidding.

Everybody knows or should know what syntax is, since he was exposed to this in the study of language in school. It relates to the form of the language, that is, the rules by which the different parts of speech are put together to form phrases, clauses and sentences. It is, as it were, the logic of language; verbs must be conjugated and nouns declined, etc. Syntactics extends this sort of thing to signs of all sorts. For example, the definitions and postulates of a mathematical theory, which set up the rules by which the entities of the theory are to relate

[9] C. W. Morris, *Signs, Language and Behavior* (New York: Prentice-Hall, 1946).

to each other, form a branch of syntactics. In fact, pure mathematics is an illustration of a language characterized only by syntactics. We shall go into this at length later.

The next step is semantics, which concerns the association of things, ideas, and in general elements of human experience with signs. When signs become symbols, we are said to be dealing with semantics. Hence, most people connect semantics with meaning, since when we relate a sign to an object of an experience, it is evidently because we think it is important, and we wish to have some unambiguous way of calling attention to it in relation to other objects designated in similar fashion. It is perhaps well to point out that information theory as developed earlier in this chapter has nothing essentially to do with semantics. It does not concern itself with the meaning of the messages transmitted but only with the recognition of the signs.

Since language does have in general a significant effect on human beings, i.e., acts as a stimulus to which the organism can make a response, we need yet another discipline to describe these phenomena. This is the function of pragmatics. Its importance may be readily recognized from the fact that two messages consisting of precisely the same signs arranged in the same order, and which syntactically and semantically are identical, may nevertheless appear to be quite different messages so far as their influence on two separate individuals is concerned. "Tom has just broken a record" means one thing to a person who is receiving information about an athletic contest and quite another to one who has been informed about an unfortunate mishap in his collection of phonograph discs. Similar examples are legion, and they occur in all circumstances of human experience. The message conveyed by the sentence: "The mass of the proton is approximately 1840 times the mass of the electron" certainly affects quite differently the man in the street and the earnest student of atomic physics, though its syntactical and semantical status is strictly speaking the same for everybody. Every scientific statement is really an illustration of the same thing. This involves the whole problem of the impact of communication on the individual human being, a matter of vital importance to educators and advertisers alike, not to mention politicians. We shall return to this in our specific discussion of the communication of scientific ideas.

At the moment, we wish to stress another aspect of language of great interest in information theory. This is its *redundancy*. It is common knowledge that nearly all of us in both speaking and writing incorporate more signals than are really essential to make the message convey the meaning we have in mind. Thus we habitually (in English) use definite and indefinite articles though often these are unnecessary. We often repeat the same set of signals, as we say, for emphasis; presumably this was originally done in the fear of being misunderstood, as in talking over the telephone, where indeed we are occasionally reduced to spelling proper names letter by letter. We seem to be torn between conflicting anxieties here. On the one hand, minimum effort dictates the use of the smallest possible number of signs per given message. On the other hand, there is no point in a message at all unless it produces in the recipient the effect we are seeking, and we must at all costs not be misunderstood. In a certain sense, language evolves in the competition between these opposing tendencies, and inevitably redundancy accompanies its development.

Redundancy in language is closely associated with what the information-theory experts refer to as "correlation." We have already seen in the previous sections that in any actual language the letters are not associated with each other in purely random fashion. Thus in English the letter t is more often followed by h than by any other single letter, because of the frequency of occurrence of the definite article "the." Then too, certain combinations of letters occur with such great frequency that if one sees the first few, the rest could be supplied by the recipient of the message with very little uncertainty. Thus when we see "oug," we feel pretty certain that h ought to occur at the end. Hence to put in the h provides a degree of redundancy, and much English spelling provides ample illustration of this tendency. The moves toward simplified spelling and "basic" English may be taken as indications of the desire to remove some of this form of redundancy. We shall return to this problem in the somewhat more elaborate discussion of the statistical study of language in the next section.

We tend to take language (at any rate our own language) so much for granted that we probably fail to appreciate its importance in all aspects of life. It is true we realize that our children do not come equipped with a command of it and must learn it. Moreover, we are

aware that real competence in its use from a cultural point of view
demands more study and practice than the vast majority of people
are willing to give. But the really marvelous power of this instrument without which social life would be impossible stirs little enthusiasm in most people. We shall not try here to assess the comparative
shortcomings of the grammarians, the journalists and the advertisers
in this business!

The place of language in culture has been the theme of countless
scholarly studies, and it would be impracticable even to summarize
their content. The problems posed are numerous and difficult. For
example, just how does language arise in the first place? We have
seen that spoken language almost certainly antedated written language, but the transition is still shrouded in mystery, though various
theories about this have been advanced. Rather, more attention is
being paid nowadays to the logical structure of language and to its
connection with the cultural activities of human beings. We might
stop to notice one interesting point of view advanced by the late
Benjamin Whorf,[10] an American anthropologist who devoted much
time to a study of the languages of the American Indians and their
relations to thought and behavior. He came to the conclusion that
contrary to a commonly-held notion that human cognitive processes,
that is, the development of our ideas, precede the expression of them
in language, the converse is actually the case: linguistic patterns determine modes of thinking and even further our ways of looking at
the world around us. This is a bold hypothesis, which some may find
rather hard to swallow, at least in so far as the early development of
human language is concerned. It is true that some authorities on language are sure that the artifacts found among the fossil remains of
primitive man (i.e., our progenitors of some million years ago) demonstrate that early man must have been able to communicate by
speech. For it is assumed with some plausibility that the production
of such materials as primitive tools (arrowheads, etc.) implies fairly
advanced cooperative behavior, and this in turn implies communication on a more or less sophisticated level. On this argument, man has
had a long, long time for his language to affect his cultural attitudes.
Yet the lingering doubt remains: did not the primitive reaction of

[10] Benjamin L. Whorf, *Collected Papers on Metalinguistics* (Dept. of State,
Washington, D.C., 1952).

early man toward his experience necessarily precede anything he said about it and was not this reaction essentially a cognitive one? The problem is a difficult one, and we cannot hope to examine here the evidence on both sides. Perhaps a reasonable viewpoint is that the relation between language and cognitive processes and attitudes toward experience is a mutual one, a case, so to speak, of action and reaction; some sort of nervous response to external stimulus produces speech sounds, and these gradually develop an orderly one-to-one correspondence with elements of experience. But as this language becomes more sophisticated, it begins to exercise an increasing influence and even control over the kind of experience considered of importance for man's well-being and the interpretation of this experience. This is especially evident in the language of science. A whole train of theorizing may develop about a key word like "energy," for example, and without question this tends to color the way we look at physical phenomena. No matter what the physical experience is, whether large-scale meteorology or small-scale nuclear physics, the attempt is made to fit the experience to the conceptual picture associated with this word. And we must not forget that all the concepts of science are expressed in words, which in turn are symbolic of operations, either in the laboratory or with pencil and paper.

We see, therefore, that even if Whorf's hypothesis may not be acceptable in its original, strong form, it suggests a relevance of language to our ways of operating which merits a lot of attention and indeed sophisticated scientific exploitation.

The Statistical Study of Language

As we have already stressed, there are many ways in which we can study language depending on whether one's interest runs to grammar, etymology, phonology, etc. But the modern scientific attitude toward language is inevitably quantitative, and this essentially implies a statistical study. We can see this in simple fashion if we decide to consider language, whether spoken or written, as a device for putting meaningful order into an otherwise random collection of signs. At once statistics enters if we are to do anything with this idea. For if we take a finite number of different signs, e.g., the letters of the alphabet, there exists a very large, though finite, number of possible combinations of these signs to make words of a given length. Thus, to take the famous four-letter word as an example; simple algebra shows that there are 456,976 ways in which one can form a four-letter word with the 26 letters of the English alphabet,[11] provided of course repetitions of letters are permitted. This means that many of the "words" would look very peculiar indeed, since among them would be such creations as aaaa, aabb, abac, etc. Most of these combinations will be meaningless in actual English, but a certain fraction of them and other letter combinations of various lengths will in the course of time come to convey information by virtue of their order. Otherwise put, certain arrangements of the letters will come to be thought of as orderly, while others are disorderly or random. This ultimate estimation is not of course purely arbitrary, since in English and other phonetic languages the relation between sound and letter arrangement serves to pick out certain combinations as more useful than others and hence more "meaningful." The word "useful" here obviously implies actual use. Hence arises the possibility of studying the frequency with which words are used. This is a statistical study, and as we have already seen, it has been carried out by Zipf, who conducted word counts on large works of literature such as novels, as well as newspaper files. In this he turned up the famous frequency-rank order relation

$$P_n = An^B \tag{1}$$

already commented on, with $B = -1$ approximately for normal English prose. (Recall that n is the rank order and P_n the frequency of occurrence of the word type.) Here then is a statistical law of language. Qualitatively, it appears obvious, but its particular quantitative form (i.e., the exponent minus one) suggests that it results from some rather fundamental mental attitudes of the human beings who construct their language to fit the needs of life.

The French mathematician B. Mandelbrot[12] decided to try to explain Zipf's law by a theory of the construction of language, based essentially on the idea that in devising language human beings are

[11] This is really just 26^4, which can be justified on the basis that there are 26 ways of picking the first letter of the word, 26 ways of picking the second, etc. The product of 26 by itself 4 times then gives the total number of possible words.
[12] Colin Cherry, *op. cit.*, *On Human Communication*, pp. 105, 209.

activated by the basic desire to economize effort as much as possible, a plausible hypothesis and one that has appealed to many theoretical sociologists, who have perhaps been influenced by the success of the principle of least action in physical science. The general course of Mandelbrot's reasoning can be described without going through the analysis, though it is a pity more people refuse to learn the language of mathematics, which possesses such power and is so economical. Perhaps the well-advertised aversion of the human race to mathematics is based on an average laziness, which may be thought to exemplify the principle of least effort. If so, it is ironically unjustified since in the long run the use of mathematical language minimizes effort, at any rate in relation to the logical difficulty of the problems it attacks and solves. But we should not digress about this here, since we shall go into it very thoroughly later in this chapter.

To return to Mandelbrot and Zipf's law, we shall after all ask the reader's indulgence for a slight excursion into a part of the mathematical analysis. The first and basic procedure of the former is to assign a *cost* to each word; that is, to assign to each word a number representing the relative effort involved in using it. In general, a word of many letters will thus possess greater cost than one of fewer letters; it is said to be *worth* more. (This of course is in a technical sense!) But a word is made up of letters, and so one has really to start by assigning a cost to each letter. Mandelbrot makes the simplifying assumption that all letters are equally costly, which normal experience certainly does not justify. However, he is able to show that the general character of his demonstration leads to the same conclusion even if he does not make this assumption. To get the idea of his procedure, it is better not to complicate matters too much, and so we cling to his original ideal assumption. Now in a language with an alphabet of M letters, we have already seen that there are M^l words containing l letters. Now the rank order of any word of length l is the sum of all words equal to or less than l in length, or if we denote this rank order by n_l we have

$$n_l = \sum_{k=1}^{l} M^k \qquad (2)$$

If M is large compared with l, this sum is approximated with sufficient accuracy by M^l (try this out!). We see that the rank order increases very rapidly with the length of the word and this is certainly reasonable; we expect very long words in general to rank far down in the list, as we arrange words in order of the frequency of their use. (Recall that n_l is a small integer for words most frequently used, etc.) We can turn this relation between n_l and M^l around and solve for l in terms of n_l, and obtain

$$l \doteq \log_M n_l \tag{3}$$

So the length of a word is the log to the base M of its rank order—approximately. But now it is a reasonable assumption that the cost of a word is directly proportional to its length. Hence the cost, which we may write C_{n_l}, becomes

$$C_{n_l} = K \log_M n_l \tag{4}$$

where K is some constant.

So far, the mathematics associated with Mandelbrot's deduction is simple algebra. Now we come to a more difficult part, namely, that which connects the frequency of occurrence of words with the cost. Here we have to form the expression for the average cost per word in actual messages. This will be the cost C_{n_l} averaged over the frequency p_{n_l} of its occurrence: words occurring more frequently will of course contribute more to the cost than those of lower frequency. Mandelbrot then carries out the mathematical operation of making this average cost per word in messages made up of words of all rank orders, a minimum subject to the condition that the rate of flow of information shall remain constant. He takes this rate of information flow as H in eq. (10), page 151, and interprets the p_i there as proportional to the word occurrence frequencies p_{n_l} already introduced. This is not a difficult mathematical operation, but we shall here merely give the result, namely

$$p_n = Pe - QC_n \tag{5}$$

where P and Q are new constants and we have left off the subscript l from n for simplicity. For the reader not familiar with the properties of the exponential function, we plot p_n as a function of C_n in Fig. 6.2. For $C_n = 0$, zero cost, the frequency p_n is a maximum and

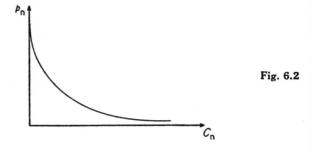

Fig. 6.2

equal to P. As C_n increases, p_n falls and approaches asymptotically to zero. The rate of decrease is directly proportional to the constant Q. The curve reflects the rather obvious fact that the less frequently used words (i.e., those for which p_n is relatively small) are the most costly ones.

The final step is the elimination of C_n between eqs. (4) and (5). This gives the desired relation between the frequency p_n and the rank order n. The reader with a little skill in algebra should have no trouble in showing that this relation is equivalent to Zipf's law in the form given in eq. (1) with

$$B = - QK/\log_e M \qquad (6)$$

Since Zipf found experimentally that for sufficiently large samples of English B is approximately minus unity, it follows from eq. (6) that in this case

$$QK = \log_e M = \log_e 26$$

Our purpose in outlining this deduction of Zipf's law has been in part to emphasize the possibility of giving a scientific interpretation of a statistical law of social behavior, but more particularly to stress the value of the statistical point of view in the study of language. The possibilities opened up by Zipf's law are very suggestive, as he himself made eminently clear. One could, for example, contrast the shape of the frequency-rank order curve for normal prose in a given language with that for scientific and technical literature or for children's books. By studying older classics, e.g., Shakespeare, one could trace the temporal change in certain aspects of language. There is no reason why the same technique should not be applied to spoken language.

We can do even more interesting things with Zipf's law. If we interpret the p_l in eq. (10), page 151, as equivalent to the p_n in Zipf's law, i.e., as the fractional frequencies of occurrence of word types, we can use the law to compute H or the average information per word in the collection of words being investigated, i.e., the amount of information needed to send the word as a message. L. Brillouin gives an example[13] taken from C. E. Shannon in which the first 8727 word types in a given text were arranged in rank order and Zipf's law reduced to the form

$$p_n = 1/10n \tag{7}$$

where p_n is a normalized fractional frequency so that

$$\sum_{n=1}^{8727} p_n = 1 \tag{8}$$

Then the average information needed per word becomes

$$H = -\Sigma p_n \log p_n$$
$$= -\sum_{n=1}^{8727} \frac{1}{10n} \log \frac{1}{10n} \tag{9}$$

We must remember that the logarithm here is to the base 2. To convert to base e we must write

$$H = -G\Sigma \frac{1}{10n} \log_e \frac{1}{10n} \tag{10}$$

where

$$G = \log_2 e = 1/\log_e 2 = 1/0.693 \tag{11}$$

This can be rewritten

$$H = G \sum_{n=1}^{8727} \frac{\log_e 10n}{10n} \tag{12}$$

According to Brillouin, the evaluation of this sum yields

$$H = 11.82 \text{ bits per word} \tag{13}$$

[13] L. Brillouin, *op. cit.*, p. 24.

This is then the information needed to send the message corresponding to the average word in the text in question. If we assume that the average English word contains about 5.5 letters, this reduces to 2.14 bits per letter.

Earlier in this chapter we developed some ideas on thermodynamics with reference to information theory. In particular, we discussed the role of entropy production and consumption as a measure of the change from order to disorder and disorder to order, respectively. It is not surprising that these ideas have been applied to the statistical analysis of language. We close this section with a reference to some recent work in this connection.[14] It must be emphasized that the work in question is a purely *structural* analysis. Thus, the investigation divides up a given text (novel, poem, philosophical treatise or whatever it may be) into structural elements and then introduces characterizations for each element. Thus, an element may be a single letter, a word or a group of words like a phrase or clause. The characteristics might be the number of letters in a word, the number of syllables, the syntactical character of the word, etc. The structural analysis then consists in studying the functional relation between the characterizing property and the number attached to the element in some sort of order. To give an example, suppose that the elements are words (really *word types* in the sense of Zipf's law) and the characterizing property is the number of syllables per word. Let the words (R in number in the given text) be tagged with the numbers 1, 2, 3, $---$, r, $---$ R, and the number of syllables be 1, 2, 3, $---$ j, $---$ J, where J is the maximum number. Then j is a function of r, i.e., each word has a definite number of syllables. Let the number of words which have j syllables be denoted by N_j. Then the relative fractional frequency of words with j syllables is

$$p_j = N_j/R \tag{14}$$

where, of course,

$$\sum_{j=1}^{J} p_j = 1, \tag{15}$$

[14] Wilhelm Fucks, "Mathematische Analyse des literarischen Stils," in *Studium Generale, 6,* 506 (1953).

since the total number of all words in the text is R. The *average* number of syllables per word is an important property of the language structure. This is given in the usual fashion by

$$\bar{j} = \sum_{j=1}^{J} j\, p_j \tag{16}$$

Fucks call this a *style-characteristic*. Another such characteristic is the Shannon measure of information given by eq. (10), page 151, where to be sure the p's now refer to word frequency rather than letter frequency. We may if we like call this measure the entropy per word with respect to the number of syllables in it. Thus

$$S = - \sum_{j=1}^{J} p_j \log p_j \tag{17}$$

In the usual information theory technique, the logarithm is to the base 2. Presumably Fucks on the other hand uses e as the base, though he does not state this explicitly. In any case, the matter is not important, since it involves only a multiplicative constant, like the G in eq. (11), above. In general, one is interested in the relative values of S for works in different languages by different authors, etc.

By word counts the following table of values for $\bar{j}$ and S has been constructed by Fucks.

TABLE 6.2

AVERAGE NUMBER OF SYLLABLES PER WORD
AND ENTROPY PER WORD FOR ENGLISH, GERMAN, GREEK, AND LATIN WORDS

		$\bar{j}$	S
Shakespeare	*Othello*	1.29	0.29
Galsworthy	*Forsyte Saga*	1.34	0.33
Huxley	*Brave New World*	1.40	0.37
Mann	*Buddenbrook*	1.74	0.51
Jaspers	*Der Philosophische Glaub*	1.89	0.50
Euripides	*Orestes*	2.13	0.59
Sallust	*Epistula II*	2.48	0.64

The analysis shows clearly an upward progression in both $\bar{j}$ and S
as one proceeds from English (the relatively constant values in English from Shakespeare to Huxley, 300 years apart, is striking) through
German to Greek and Latin. It takes simpler words, on the average
more monosyllabic in character, to express messages in English than
in the other languages studied. At the same time, the entropy, measuring the mean information rate per word, is least for English and
greatest for Latin, though Greek is practically on a par with Latin.
This means that the amount of information per word as technically
defined in this chapter to convey messages in English words is decidedly less than in the other languages. Correspondingly, the entropy per word is lower.

Several observations may be made about these results. In the first
place, it might be remarked that they could actually be obtained
by a very cursory examination of the texts in question, at least so
far as $\bar{j}$ is concerned. Everyone who has studied Latin or Greek at
all knows the high degree of inflection of these languages. This
means prefixes and suffixes and adds to the average number of syllables. And then, so far as entropy is concerned, what good does it
do to know that the entropy of English prose and poetry is less per
word than that of German? One must answer candidly that the
application of the concept of entropy to this matter is by no means
necessary. Indeed, some authorities on the subject of communication,
for example Colin Cherry, have grave doubts about its ultimate
value. On the other hand, it does provide a formal objective criterion for the comparison of different languages, previously overlooked by linguists, and it is at any rate worth further exploration
if indeed we wish to introduce quantitative considerations into what
has hitherto been handled largely in qualitative fashion. Thus, if
one plots the entropy S as a function of the average number of
syllables per word, $\bar{j}$, one gets for the various languages so far studied
a distribution as shown in Fig. 6.3. All works so far studied have their
$S, \bar{j}$ values lying more or less in the cross-hatched region. Inevitably,
a relation of this sort raises the question, why? It also suggests study
of material in a great many other languages to see where their corresponding points lie. Of course we have taken only a very simple
example. Other language elements could be studied and with more

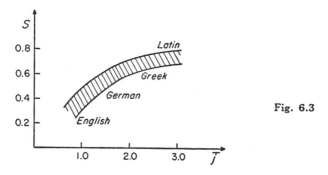

Fig. 6.3

From Wilhelm Fucks, "Mathematische Analyse des Literarischen Stils," Studium Generale, 6, 506, 1953.

complicated properties than the number of syllables per word or the number of words per clause. The possibilities are endless, and we may be sure that someone is bound ultimately to explore them without worrying to start with about the possible value of the results. One possibility to be sure occurs at once. This is the application to the problem of settling the authorship of doubtful texts from antiquity. For details on this, the paper of Fucks should be consulted.

The Communication of Scientific Ideas

It is proper to reiterate what was said at the beginning of this chapter: without communication there could be no science. Each potential scientist in every generation is brought up on the published accounts of the results of scientific research in previous generations, either the original writings preserved in the archival journals and proceedings of learned societies or more commonly (in the early stages of education in science) textbooks summarizing the state of the various sciences at the time. In either case, communication is involved, and the success of the training of the embryonic scientist in each age depends on the skill and completeness of the communication process in the material being used. The same can be said for the oral presentation of the science teacher, who is supposed to supplement the "book" and make it "clear." He is supposed to do more: in addition to making science intelligible to the young, he must make it seem to be a worthwhile pursuit; young people do not normally decide to devote themselves to a profession merely because they think they understand what it is about; they must find it interesting as well.

But the communication of scientific ideas is not of importance only to the person intending to become a professional scientist. If science and its applications really play the role in civilization that is now freely admitted on all sides, the obligation to communicate something about science to the general public follows as a matter of course. Otherwise, science becomes a new form of magic to the man in the street and a black one at that. It may be said, "What difference does it make if the citizen knows nothing about thermodynamics and electricity as long as the car starts when he turns the ignition key and puts his foot down on the accelerator?" Put thus bluntly, there is a certain sense in which it *does* make no difference. The trouble is that an ever-increasing fraction of the money the citizen spends for everything he buys is going to the support of scientific research which industry feels is at the basis of the improved technology of the future. Moreover, more and more of the citizen's tax payments are being devoted to government support of scientific research. He would be an idiot indeed if he did not feel he ought to know something of what goes on in this business. It is the obligation of society to tell him, and this is a problem in communication.

Since this kind of communication to be effective must begin with the young, we are really talking about a problem in education. The history of science education in schools and universities throughout the world has not been by and large a happy one, though the pattern has varied somewhat from country to country and age to age. The teaching of science in the elementary and grammar school grades (pre-high school) in the United States was largely confined to simple nature study until very recently, and the situation has not been much better in the European countries. Secondary-school science teaching in physics, chemistry, biology and astronomy has been by and large uninspired in this country and Great Britain, and this has persisted to the present, though great efforts are now being made to put life and meaning into it. Speaking broadly, it is only during the past century that the universities of the English-speaking countries have paid much attention to the teaching of science (outside of mathematics) as compared with the humanities. There have been notable exceptions, of course, as in the Scottish universities and in certain American technical institutions like Rensselaer Polytechnic Institute. It is a striking fact connected with this situation that great discoveries in science in these countries before 1860 tended to be made by investigators not connected with universities. We think, for example, of British scientists like Young, Faraday, Joule, Davy, Darwin, etc. The famous Cavendish Laboratory at Cambridge University under Maxwell was not founded until 1870. Anyone who wishes to get a picture of the struggle in England to get science recognized as an appropriate subject of collegiate instruction should read the writings of Thomas Henry Huxley,[15] who himself played such a dominant role in overcoming the opposition of the obscurantists who felt that science had nothing to do with a gentleman's education.

The battle for the acceptance of science as a suitable subject for instruction in schools, colleges and universities in the English-speaking countries has of course been fought and won; however, this has not solved the problem of adequate education in science for all who go through these institutions.

We need not linger long over the educational procedures for those who intend to become professional scientists, though these are worth a few comments, because we need not pretend that the professionals in the universities are even now in possession of the whole wisdom in this matter. It might indeed be said that if the professional academic scientists in a given field, say physics, are not competent to prescribe the correct course of study for the young people who will be their successors in the profession, who is? The difficulty lies in the well-known increasing fragmentation of learning in every discipline coupled with the inevitable development of modes of interest which tend to encourage large groups to specialize in what is fashionable at the time. This has an obvious impact on the professional course of study, particularly at the graduate school level. Nuclear and solid-state physicists, for example, who are now in the saddle in most university physics departments, prefer to stress preparation in particle mechanics and electromagnetic theory and are apt to pay little attention to thermodynamics and continuum mechanics.

[15] See, in particular, T. H. Huxley, *Science and Education—Essays* (New York: D. Appleton, 1902).

Hence, different generations of professional physicists are apt to come up with rather different backgrounds. Perhaps this is after all not so serious. Rather more of a problem is the tendency on the part of professional scientists (or professionals in any field for that matter) to neglect breadth and hence to miss the important relations between their own field of specialization and other ways of looking at human experience. Special efforts should be made to see that every professional scientist becomes acquainted in some measure with the history and philosophy of science, and through these means, as well as others, establishes a rapport with the humanities. It must be confessed, indeed, that the need for this is in general by no means so great as the corresponding need for appreciation of the significance of science on the part of the specialists in the humanities. For scientists, being human, can hardly escape, even if they wished to do so, contact with general human affairs through newspapers, magazines, radio, and mass media in general. On the other hand, it is all too easy for the nonscientist to remain free of contact with science except through the factual impact of its technological applications.

This brings us to general education for the nonscientist. Here the problem of communication has deeper roots than the mere transmission of scientific ideas as such. Before the student can learn anything about science, he must have achieved some mastery of the techniques of communication generally; thus, he must have learned his own mother tongue so as to be able to speak it, read it, and write it with fluency. There is scarcely any more vital prerequisite for understanding of science than this, which of course is essential for an ability to grasp any aspect of human experience. But since the language of science naturally must be more precise than that of common everyday living, ability to communicate and receive communication in the mother tongue is all the more important if both the spirit and content of science are to be understood. The student whose grasp of language is so feeble that he believes, for example, that the word "force" always means the same thing wherever he finds it will find it difficult to understand that its use in physics has nothing to do with "police force."

A word about mathematics must be uttered here. If the language of ordinary speech provides the instrument for qualitative understanding, mathematics is *the* quantitative language and indispensable for science. It is without qualification the most marvelous language ever created by the mind of man. The true international language, it should be part of the cultural equipment of every thinking being and particularly for all who desire to develop an appreciation of science. It used to be thought that this was true only of the physical sciences and that the life sciences would never be "contaminated" by mathematical analysis. But the urge to figure is irresistible among scientists, and biology is using more mathematics all the time. The same is true of psychology and the social studies, as has been illustrated in previous parts of this book. In referring to mathematics here we must be careful not to restrict our attention to the purely metrical aspects, overwhelmingly important as these are. Not to be overlooked are the more profound logical implications, the algebra of relations and those considerations which have come to be called topological. These will play an increasingly significant role in contemporary and future science.

So much for techniques! What can we say about instruction in the various branches of science themselves? Here it seems that the watchword is *curiosity,* the very root of science and one of the prime and apparently inborn characteristics of children. The young person who does not have this trait encouraged at every possible opportunity will never know what science is about, even though he may be able to parrot a few memorized details. Hence, general education in science must capitalize on curiosity; this is true indeed for children of all ages. One of the melancholy features of much elementary education, and even the advanced variety for that matter, is the tendency to beat the inquisitive urge out of the student. This is fatal to the endeavor to promote a genuine understanding and appreciation of science and of most other human intellectual activity as well. Some may feel that to stimulate curiosity about all features of human experience does not comport with the inculcation of technical skills. But surely the educationists, who are forever busy devising new methods of instruction, should not find this problem beyond their powers.

Science instruction should take place at every level of schooling, beginning with kindergarten. For very young children it should take the form of encouraging native interest in the external world and developing powers of observation and description. Older children should be introduced (this can be done with eight-year-olds and up) to the idea of the experiment; they should be shown some involving very simple equipment and urged to try variations on their own initiative. Later, the fundamental significance of experiment as a form of creation of experience should be discussed and the history of some famous experiments should be recounted with emphasis on their relevance for the development of certain scientific concepts, e.g., energy. At all costs, the presentation of science as a mere collection of facts should be avoided. To this end in secondary school and college, experiments should preferably be of the project variety rather than the "cookbook" affairs that are typical of present-day curricula in general education in science. In any case, emphasis should be laid on the need for the individual to test things for himself and not to rely merely on authority. If a textbook is used, it should be for the purpose of questioning every statement in it before it is accepted as valid and meaningful.

All this of course puts quite a strain on the teacher. Who after all is going to be the appropriate person to communicate an appreciation of science to the students of our schools, colleges and universities? His is obviously a position of considerable responsibility, and it does not seem possible for him to discharge his duties adequately unless he is essentially a learned individual. Here it is clear that we must aim at the highest possible ideal and not be deterred by the very real obstacles in the way. We could do worse than take as our guide one of the greatest humanistic scholars who ever lived, Desiderius Erasmus of Rotterdam, who once wrote to his young friend Christianus of Lübeck about the proper method of study: "Let it be your first care to choose you a master who is a man of learning; for it cannot be that one that is unlearned himself can render another learned." This clearly means that the teacher of science must possess a deep understanding of his subject and not be merely a master of pedagogy. It is all too common to emphasize that most scientists whether they be teachers or not are specialists and "know" only their own fields. But this is a myth disseminated only by those who are unable to distinguish between the fundamentally important elements of science, namely, its logical structure, concepts and principles on the one hand and the detailed facts which can be found in all the handbooks.

The successful teacher of science must have caught the spirit of research and his training should therefore preferably include some research. Otherwise, how can he hope to instill in his students honest feeling for what is the lifeblood of science? One of the most clear-cut statements of this need was made not by a scientist but by a distinguished Oxford humanist, Mark Pattison, who wrote in 1868:

> No teacher who is a teacher only, and not himself a daily student, who does not speak from the love and faith of a habitual intuition, can be competent to treat any of the higher parts of any moral or speculative science.

Going on to refer to the lamentable situation then prevailing in many departments of the British universities, Pattison continued:

> Our weakness of late years has been that we have not felt this;— we have known no higher level of knowledge than so much as sufficed for teaching. Hence education among us has sunk into a trade, and like trading sophists, we have not cared to keep on hand a larger stock than we could dispose of in the season.

Perhaps Pattison was unduly severe on his contemporaries. But the danger is a very present one, that our teachers of science shall fail to be in some measure scholars in science and willing and able to devote themselves to lives of scholarship. But it is only in this way that the process of communication from teacher to student and the even more important communication from scientifically described experience to student can take place with any effectiveness.

Though the education of the young is the key factor in the problem of the communication of science to the populace, we must not ignore the possible influence on the adult of popular science literature. To assess the impact of this is naturally a bit tricky, since it is so completely mixed up with the general hodgepodge of mass media and so frequently misleading if not downright wrong. One of the obvious difficulties is the well-known confusion between science and technology. We go further into this in the next chapter and desire here to take no pejorative attitude toward technology as such. But it is important that the general public should be able to distinguish it from science. It is obvious that the popularizers of science should have a thorough understanding of the relation in question. They must also be in a position to make clear the conceptual character of science and avoid the temptation of forever dishing out a pot-pourri of miscellaneous facts. The task is not an easy one. Perhaps the more extensive scientific training of a class of journalists is one answer. Even better would be the collaboration of eminent scientists, who know what science is and what they are talking about, with competent journalists. There seems to be a rather widespread opinion among newspapermen that scientists cannot write about science in a way that is attractive to the average newspaper reader. Though professional technical articles and books are naturally too formidable for the general reader, there seems no valid evidence that the opinion just mentioned is justified. In fact, given sufficient encouragement, scientists can describe their ideas in simple language, and there exist many examples. In English-speaking countries, it is true, the British scientists have in general excelled the American in both their willingness to instruct the general public and their skill in doing it effectively. But here again success has been closely tied to the effort to promote genuine understanding rather than to provide "exciting" facts. We need to revive and extend the tradition set by the great popularizers of the past who were also great scientists, men like Faraday, Huxley, Tyndall, Jeans and Eddington in the English-speaking world and Helmholtz, Mach and Poincaré on the continent of Europe. The new popularizer must make clear that science is tied not only to technology but also to the humanities. He must be willing to take advantage of every communication outlet, including not merely the printed word and picture but also radio, television, movies, public lectures, museum displays, etc. All these techniques actually are now being used, but the magnitude of the operation must be vastly increased and more imaginatively handled if wrong impressions about the role of science in civilization are to be replaced by appropriate ones in the minds of the general public.

Mathematics as the Language of Science

The striking thing about the position of mathematics in science is its ambivalence. It plays the dual role of a branch of science in its own right and also of the language of science. In the usual dictionary definition it does indeed figure as a science; we have in Chapter 2 used geometry as an appropriate example of a scientific theory. To be sure, the elements of experience which mathematics in this sense seeks to describe and understand are idealized entities and not closely tied to operations immediately evident to the senses or performed in a laboratory. Nevertheless, the marks which the mathematician puts down on paper and to the relations among which he assigns arbitrary but highly precise rules are definitely elements of human experience, and in operating with them the mathematician goes through the same logical scheme outlined for scientific theories in general. There is, to be sure, one exception: he has no way in which to carry out the last step, namely, the experimental test of the mathematical "laws," i.e., the theorems resulting from his theory. All he can do is to compare the conclusions with the hypotheses and all other conclusions drawn from the same assumptions to see whether they are all consistent with each other. The absence of logical contradiction is his only assurance of the scientific validity of his work. Of course, he can, if he chooses, assign operational significance (in the natural science sense) to his symbols and translate his theory into one recognizable as scientific and amenable to the usual laboratory or observational testing. When he does this, we are apt to call him an *applied* mathematician. The so-called *pure* mathematician will disdain this recourse. He has no interest whatever in giving operational meaning to the symbols, save in terms of purely mental or pencil and paper operations, e.g., addition and subtraction, multiplication and division, etc.

Thinking of the pure mathematician reminds us of Bertrand Russell's famous definition: mathematics is the class of all propositions of the form "p implies q." This seemingly reduces the subject to deductive logic. At the same time, we must not forget that the same philosopher also defined mathematics as the subject in which we never know what we are talking about nor whether what we are saying is true. The latter is certainly enough to startle those who have been accustomed to look upon the subject as the most precise of the sciences and the repository of all guaranteed truth. A little examination clarifies the situation and removes some of the astonishment. For, since the pure mathematician operates merely with marks on paper, which have no connection with human sensory experience, the outside observer may be pardoned for feeling that the mathematician really does not know what he is talking about: the counters in his game are not stones, flowers, animals or atoms; they are meaningless marks symbolic of nothing. Or so it seems! And since the postulates set up to govern the behavior of the chosen entities are completely arbitrary, we have no assurance of any "truth" value to be associated with the deductions from them. Now a little thought will throw considerable doubt on the value of Russell's definition, except to give us a bit of a jolt. Though the mathematician's counters, his x's and y's and a's and b's, appear to be rather insubstantial to the scientist who gives symbolic value to what he calls actual physical objects, the difference is probably illusory. In fact, once the mathematician has assigned relational properties to his entities, he certainly knows what he is talking about when he uses them. Since he has never had to worry about whether they correspond to anything else in his experience, he may well be justified in feeling surer of his grip on them than, say, the physicist who has to make a pretty big extrapolation from his experiments to his conceptual atom. At the same time, whether the statements of the mathematician are "true" or not in any fundamental philosophical sense has little special relevance to mathematics as a science, since precisely the same problem arises in other branches of science.

We conclude that Russell's second definition is a stunt. The first definition which makes mathematics a branch of logic commands greater attention and respect. Every thinking person in our Western civilization aspires to reason logically, i.e., arrange his thoughts in conformity to certain arbitrary but apparently plausible rules, which have long been accepted as a part of social convention. When the logical processes are expressed not in terms of ordinary colloquial speech but in terms of a set of abstract signs, the gain in economy and accuracy of thinking may be considerable. As a very simple illustration, consider the common syllogism:

> All men are mortal.
> Socrates is a man.
> Therefore Socrates is mortal.

It is readily recognized that the type of reasoning involved here can be expressed more generally in terms of abstract symbols. Let anyone in the class of men be denoted by A and anyone in the class of mortals be denoted by B. Let the man Socrates be X. Then we may rephrase the syllogism:

All A are included in B.
X belongs to A.
Therefore X belongs to B.

Even this formulation can be replaced by a more economical one, which also has more general application. We can agree to represent by the juxtaposition (logical "product") AB the class of elements which are common to both A and B. The syllogism then takes the form

$$A = AB.$$
$$X = XA.$$
$$\therefore X = XAB = XB.$$

The operations involved should not be confused with those of conventional arithmetic and algebra. Here AB is not the actual algebraic product but the logical product. The sign $=$ here means "is the equivalent of" or "is logically identical to." In any statement we are entitled to replace the symbol of any class by a logically equivalent one. With these assumptions the above formalism becomes much more powerful and general than the verbal representation. This illustrates in a very elementary case the power of abstract symbolism in logic. Actually, in modern logic it is more customary to employ the symbol $\subset$ to mean "is included in," so that $A \subset B$ means that the class A is included in the class B, etc. However, it is not our intention to provide an introduction to modern symbolic logic.[16] Bertrand Russell and A. N. Whitehead endeavored to show indeed that all mathematics is but an extension of logic.[17] It is not yet agreed among mathematicians that this program has been achieved. If it were, it would have the interesting result that anyone who could think logically could also understand and use mathematics; in fact, would be doing it every time he carried through a logical argument, even if he shunned the manipulation of abstract signs. It is believed by some psychologists that there are some persons who are constitutionally unable to think mathematically. If

[16] Cf. A. Tarski, *Introduction to Logic and the Methodology of the Deductive Sciences*, 2d ed. (London: Oxford University Press, 1946).

[17] Cf. A. N. Whitehead and Bertrand Russell, *Principia Mathematica*, Cambridge University Press, vol. 1, 2d ed., 1925; vol. 2, 2d ed., 1927. Or see for a simple analysis and critical comments, R. L. Wilder, *Introduction to the Foundations of Mathematics* (New York: John Wiley and Sons, Inc., 1952).

mathematics is indeed logic and the extension thereof, this would have to be interpreted to mean that there are some persons who cannot think logically. Probably this is true, but it hardly seems possible that this is what the psychologists mean. It may well be that people who think logically when they are forced by circumstances to do so are too lazy to be willing to learn the technique for doing the job economically and with minimum likelihood of error. For learning to manipulate with symbols takes practice and can be tedious. In this respect, it is like many games.

We have prolonged our attention to mathematics, because even when considered as a science, it seems to possess such a decided communicational aspect that it is scarcely possible to overemphasize its significance for science. But this brings us to its character as a language peculiarly suitable for the expression of scientific results. This appropriateness has indeed colored much of the discussion of the previous chapters. Over and over again we have seen the power of mathematical expression in providing economy and precision. To what is this due? It is really no mystery since it emerges from the fact that mathematics is a language characterized by syntactic quality alone. Semantics and pragmatics are foreign to it, save as provided by the particular science in which it is used. Colloquial language must have all three properties, but in mathematics the symbols initially have no meaning except in their relation to each other and the various operations which they are permitted postulationally to suffer. The great miracle, and perhaps we ought to call it that, resides in the fact that with symbols obeying but a small number of relational rules (i.e., those of common algebra) we can represent such a wide variety of phenomena in human experience merely by the clever identification of symbols with the results of laboratory operations or experimental observations. No wonder so many of the physical scientists of the seventeenth, eighteenth and nineteenth centuries were carried away by their enthusiasm for the power of mathematics, which seemed to express so much in terms of such small packages, and whose manipulation made possible the extraordinarily rapid deduction of a host of conclusions from any theory whose postulates can be expressed in the symbolism of mathematics. Much of the mathematics we use today was devised indeed by men who, like Newton, were primarily interested in the scientific description of experience. And much of modern mathematics treated as a science developed from the efforts of the so-called pure mathematicians to introduce logical rigor into the useful mathematics created by scientists. A very good relatively recent example is provided by Oliver Heaviside, who developed a brand of mathematics called operator analysis, which he found extremely useful in the solution of problems involving electromagnetic oscillations and radiation. The pure mathematicians developed this into a whole new branch of mathematics.

In our eulogism of mathematics we must not overlook the fact that many snags have been encountered in its evolution as a science, and these conceivably could have repercussions on its use as a language. These pitfalls are curiously enough connected with the logical character of the subject. One such is a famous paradox of Bertrand Russell, associated with the use of the principle of the excluded middle class. Recall that the famous laws of Aristotelian or classical logic (sometimes called the "laws of thought") are: (1) the law of *identity:* every entity is what it is and maintains its identity throughout any discourse; (2) the law of *contradiction:* a particular entity cannot both be and not be so and so (i.e., of a given thing it cannot be said to be both A and not A, where A is a given quality); (3) the law of the *excluded middle class:* a given entity either is or is not so and so (i.e., again, if A is a given quality, an entity must either be A or not A; any object, for example, must either be white or not white). These laws are employed in all ordinary logical reasoning, even by people who have never realized their existence. The student of elementary geometry who proves a theorem by the so-called *reductio ad absurdum* method is really utilizing the law of the excluded middle class.

Now consider with Russell the case of the barber in the faraway land, who shaves all those and only those who do not shave themselves. This appears to be an innocent division of labor until we ask ourselves the question: does the barber shave himself? To answer the question it is natural to begin with some assumption and follow through to the consequences. Let us first suppose that the barber shaves himself. Then by the hypothesis he belongs in the class of those who do not shave themselves, since only those who do not shave themselves are shaved by the barber. But it follows then that he cannot shave himself, which contradicts the original assumption. Let us then assume that the barber does *not* shave himself. The initial hypothesis says that all those who do not shave themselves are shaved by the barber. Hence it follows that the barber must shave himself, which is again a contradiction. Hence the barber neither shaves himself nor does not shave himself, thus violating the third law of thought as well as the second law. In Aristotelian logic he is an "illogical" barber.

This is an amusing example of a rather deep-seated logical difficulty about classes of entities which are themselves classes and where we have to face the question, which seems at first harmless enough, is the class of all the classes a member of itself? Consider, for example, the class of all abstract ideas. This class itself is an abstract idea and hence is a member of itself. It may be called an extraordinary class as distinct from an ordinary class, which does not have itself as one of its elements. Thus, the class of all natural numbers, the integers, is an ordinary class, since the class is not itself an integer and hence not a member of itself. We can readily get ourselves into a logical tangle by proceeding to consider the *class* of *all ordinary classes,* which we may denote by C. Then we ask the question, is C an ordinary class? We first try to prove that it is by assuming it is not and then see if we can establish a contradiction. It is not difficult to do this, for if C is an extraordinary class, it must by definition be a member of itself. But we started out by saying that C is the class of all ordinary classes. Hence every element of C must be ordinary, and hence, if it is a member of itself, it must be ordinary. Thus, starting with the assumption that C is extraordinary, we reach the conclusion that it is ordinary; hence a contradiction. The reader should be able to show that if, on the other hand, we begin with the assumption that C is an ordinary class, we can establish logically as a conclusion that it must be extraordinary. The line of proof is, of course, precisely the same as that used in the "barber" demonstration. Hence we have to conclude that C is both ordinary and not ordinary, and this violates the law of contradiction. The only solution of this antinomy, as it is called, is either to throw out the law of the excluded middle and assume a three-valued logic, in which there is a third choice between a thing being either so or not so, or exclude such classes as the class of all classes.

We shall not try to go into the solution of this and other like antinomies, as this would take us too far into modern logic and mathematics. We have mentioned the matter only to show that even in pure mathematics "logical rigor" can become in certain cases illusory. This has led to a great deal of discussion among mathematical logicians, some of whom, the so-called formalists (of whom the German D. Hilbert has been the chief exponent), have tried to carry out a program in the standard axiomatic fashion and to establish that in any such mathematical system no inconsistency can ever be detected. Others, the so-called intuitionists, of whom the Dutch mathematician L. E. J. Brouwer has been a leading apostle, have concluded that the formalistic program is strictly impossible, largely because of a famous theorem due to K. Gödel, the Hungarian mathematician. We shall not attempt to state the theorem precisely, but its general tenor is as follows. Suppose we have a mathematical system constructed according to the formalism described in Chapter 2 with its defined entities, postulates and deduced theorems. This might be, for example, a formalized theory of numbers, or more elaborately, the system Whitehead and Russell worked out in their *Principia Mathematica*. What Gödel demonstrates is essentially that there exist theorems in such a system which can neither be proved nor disproved *within* the system, that is, by means of the postulates and deductions of the system itself. To decide such theorems it would be necessary to call upon results which are not included within the system itself. This has the essential consequence that there is no such thing as a *complete* mathematical system.

All this has naturally created a considerable stir in mathematical circles, leading to discussion of such things as metamathematics, which serves somewhat the same purpose as the metalanguages mentioned earlier. So far as science in general is concerned, it naturally raises the question: what effect will the logical difficulties about mathematics be likely to have on its use as the language of science? One might take a gloomy view about this, for it is naturally frustrating to have to think that one is unable to think straight, even when using the most powerful analytical methods which logic has provided to man. However, it is unlikely that most scientists are going to be highly exercised over this matter. The natural scientist is apt to take the pragmatic view that he does not have to worry too much about the fine points of the logical consistency of the mathematical language he uses, since in any case he can always compare his results with experience directly; a recourse the pure mathematician does not have. It is possible indeed to push this point of view too far, and as some distinguished scientists (e.g., Erwin Schrödinger) have complained, it has been too often employed to justify the use of what may be called "poor" mathematics in place of "pure" mathematics. An example would be the deliberate extension of theorems proved by mathematicians for *finite* sets of entities to *infinite* sets without concern over the mathematical validity of the extension. Some physicists do arouse the ire of pure mathematicians by their cavalier attitude. But on the whole, what may be called the "common sense" use of mathematics by scientists has worked pretty well and will doubtless continue to do so. Often, the scientist introduces a rough-and-ready crude notion like the so-called δ (delta) function of Dirac, for example, a wholly improper function from the mathematical standpoint, and is able to show that it can be used without getting into serious logical difficulties. What happens then usually is that the pure mathematicians get busy and "rigorize" the new development, just as took place with the Heaviside calculus mentioned earlier.

This suggests that we say a word, in concluding this section, about the distinction mentioned at the outset between pure and applied mathematics and its possible significance for the future of science. Applied mathematics finds its origin in the need for certain analytical methods in the solution of scientific problems. Thus, historically, the differential and integral calculus was invented to solve problems like the motion of the heavenly bodies and the shapes assumed by rotating bodies. Once devised, the method was extended to the setting up of differential equations adequate to describe the motions of continuous media like vibrating strings and columns of fluid (organ pipes, for example), and to the discovery of means for solving such equations in order to see whether by the proper identification of symbols the solutions could serve as laws for the phenomena in question. Much of the mathematics of the seventeenth and eighteenth and even early nineteenth centuries was of this character. It went along with the enormous flowering of experimental science during that period. Of course, it was accompanied by much interest in the mathematics for its own sake, as exemplified by the work of men like Euler, Lagrange, and later Gauss, Cauchy and Riemann. The nineteenth century saw a great spurt of interest in pure mathematics, with the attempt to put the calculus on a rigorous basis and

with the invention of new types of algebra and extensions of analytical geometry. This developed into the so-called pure mathematics
of the contemporary period, a science in its own right, as we have
explained, and in most cases seemingly utterly remote from application to any of the natural sciences. Some mathematicians, like the
Englishman G. H. Hardy, have indeed boasted of the lack of practical use of pure mathematics as being its strongest attraction. Many
pure mathematicians act as if they wished to have nothing to do
with applied mathematicians, not to mention other scientists like
chemists, biologists and physicists. And many scientists have recipro-
cated by calling the professional papers in pure mathematics utterly
unintelligible. They also have tended to criticize ("scoff at" might be
a more expressive term, though a bit strong) the pure mathematicians over the presumed tendency to pick fields and methods of study
which allow the relatively easy proving of a host of theorems of
somewhat specialized character rather than concentrating on really
tough mathematical problems, e.g., the four-color map problem,
Fermat's last theorem, etc. It is indeed a fact that whereas the applied mathematician has to face and do something with the problem
that is presented to him by the scientist and cannot dodge its difficulties, the pure mathematician *can* exercise considerable choice in
the problems he tackles and can thereby avoid certain difficulties if
he chooses. Many undoubtedly do so. But the criticism is probably
on net balance an unfair one, since just as the really great scientists
do not shrink from tackling the tough problems (e.g., Planck and the
black body radiation law, which led to the invention of the quantum
theory), so the really great mathematicians do not hesitate to try to
master difficult mathematical problems (e.g., Gauss, who gave the
first satisfactory proof of the fundamental theorem of algebra, that
every algebraic equation of the nth degree has exactly n roots).

There is a subtle irony about the relations between pure and applied mathematics. This is well brought out by the obvious question:
which of the two brands is more likely to provide the necessary powerful language for future science? A little excursion into history will
throw light on this. Two of the well-known British mathematicians
of the nineteenth century were Arthur Cayley and George Biddell
Airy. The latter was an applied mathematician with a strong interest
in astronomy. In fact, he became Astronomer Royal and made notable contributions to celestial mechanics. The former was a pure

mathematician and Sadlerian professor of pure mathematics at Cambridge University from 1863-1895. As such, he had much to do with the kind of mathematics taught at Cambridge. Now his most important research specialty was a new and somewhat esoteric brand of algebra called matrix calculus. At the time it seemed of no earthly use, though Cayley amused himself with the publication of several hundred papers on the subject. Sir George Airy protested against mathematical instruction of this kind, when it was obvious (to him and other applied mathematicians of his time) that the theory of differential equations was the really useful branch of mathematics, as indeed it was for celestial mechanics and the whole of the physics of that age. But Cayley kept serenely on! The interesting question now presents itself: of the two men representing, as it were, respectively the pure and applied mathematics of their time, which has made the greater impact on contemporary science? The obvious answer would be Airy, but this is wrong. Cayley's matrices proved to be precisely the appropriate language for much of twentieth-century science, including such diverse fields as electric oscillations, quantum mechanics and information theory. In other words, the pure mathematics of the nineteenth century turned out to be the most valuable applied mathematics for the twentieth.

This is by no means an isolated example. The moral of this and other stories like it appears to be that the cultivation of pure mathematics for its own sake without thought of possible application is to be encouraged for the sake of the future of science as a whole. This does not mean that every branch of pure mathematics is bound to make an impact on science within any given number of years; the point is we can never be sure in advance which are to be the scientifically fruitful ones. The only safe plan is to encourage all. Much the same argument is valid for the support of basic scientific research of all sorts. But mathematics occupies a position of the highest strategic significance from the very fact that it has provided and will continue to provide the language in which science can carry out its activities and communicate itself to all properly educated people. In this sense, we may without exaggeration say that the success of future science lies in the hands of the pure mathematicians, and express the sincere hope that they will flourish, even though they often appear to unprejudiced outsiders as a rather queer lot.

CHAPTER 7

SCIENCE AND TECHNOLOGY

The Setting of the Problem

In beginning our account of the relation between science and technology and its connection with the influence of science on society, we must once more review briefly our understanding of these two human activities. If science is a method for the description, creation and understanding of human experience, technology may be defined as human activity directed toward the satisfaction of human needs (real or imagined) by the more effective use of man's environment. Thus the provision of food and shelter counts as part of fundamental technology, even when carried out by the crudest of implements devised by primitive peoples. However, it has become customary to consider the subject from the standpoint of the tools and other machinery which have gradually been invented and improved upon for carrying out activity of this kind. The story of technology is the story of man's ceaseless striving to extract what he calls a better living out of his surroundings, and the tale includes the invention of numberless gadgets to make life at once more comfortable and more exciting. Unlike science, technology is not essentially a contemplative activity, except in so far as the inventive genius finds it advantageous to ponder on past experience in order to improve a tool or make a better one. His pondering is always directed toward a rather well-defined practical end and is not contemplation for its own sake. Yet it is obvious that in the making of gadgets, questions are bound to arise. In particular, the inventor or his observers cannot help ultimately asking themselves just why the gadgets work the way they do. When this happens, technology has initiated science. We shall give illustrations of this in the next section. This is important since it would give a wrong impression to insist that all science arose out of simple disinterested curiosity about passive experience. Early man not only looked up at the sky and asked himself why the stars move as they are observed to do; he also wanted to know why a machine like the lever could so act as to reduce the burden on a human or animal laborer.

In the sections of this chapter we are going to explore the relations between science and technology from the standpoint of both the early influence of technology on science and the later enormous impact of science on technology. This will involve a consideration of the vanishing frontier between them in the industrial research of our own times. Finally, we plan to comment on the role of technology in contemporary society, the problems this has created, and the responsibility which science, through technology, must assume in this connection.

Technology in the Prescientific Stage

We repeat the obvious fact that technology is older than science. We treat this as obvious, since all the records of early man indicate clearly that he created artifacts and evidently used them before he attempted any systematic account of the meaning of his activities. What went on in this respect before the development of written language (itself an important technological development) we can learn only through the mythology which was handed down orally at first and only much later was embalmed in writing. The relation between the mythology of primitive man and the development of science is undoubtedly a subject as fascinating as it is uncertain, but it belongs to the professional history of science, and we cannot go into it here.

However, we can remind ourselves of early technology: the discovery and taming of fire for the production of warmth and the cooking of food, the hunting of animals with the devising of weapons and tools, the sowing of seed and the cultivation of crops, the domesticating of animals, the building of shelters for the living and tombs for the dead, the manufacture of cooking and other household utensils, the progress from the age of stone as a material for the making of artifacts to the age of metal through the discovery of the treatment of ores found in the earth's crust. Finally, we must not forget those miraculous breakthroughs, which seem all too simple to us in our stage of the game but which were so world-shaking as to justify ancestor-worship if anything could, the invention of the wheel and the simple machines, the inclined plane, the lever, the pulley, and the bow and arrow. Their origin is shrouded in the mists of antiquity; their influence on the ability of early man to cope with his environment incalculable. Without them, the building of the pyramids would have been impossible; the efficient transmission of energy by convective process was enormously increased when the hurled spear was replaced by the arrow sped by the bow, making hunting a less grisly process and of course immensely increasing the "fun" of military activity.

It has not escaped historians indeed that much of the history of technology, particularly in early times, is connected with the desire of man to make war more effectively, with greater safety to himself and great devastation of his enemies. From time immemorial, people have been throwing things at other people, and a large part of technology has over the centuries been directly concerned with the creation of even more fiendishly ingenious ways of carrying out this process, which we have euphemistically referred to as the convective transmission of energy. Ample illustrations will occur in this and the next chapter.

Let us, however, get back to the mysterious properties of machines, for they must have seemed mysterious to our remote ancestors. Man early learned the social significance of the fact that human life was impossible without someone's labor, and a large part of the history of the race is the story of the efforts of the clever people to pass on this back-breaking work to others less clever than themselves, first to other human beings in the shape of slaves and then to domesticated animals. Eventually, the clever ones discovered the possibility of diminishing the effort involved in labor by the use of those devices we have mentioned, namely, machines. We ought perhaps to emphasize that a machine in its strict technological significance is a solid object or combination of solids and fluids of such a character that a force applied at one point enables a different force to be exerted at another point. Thus, the simple lever is a solid bar placed on an edge or fulcrum about which it can rotate freely. When a downward force is applied at one end, an upward force is exerted on any object placed at the other end, equal to the applied force multiplied by the ratio of the distance from the applied force end to the fulcrum to the distance from the exerted force end to the fulcrum. By choosing the distances appropriately, the force exerted can be many times that applied, yielding what is termed technically a *mechanical advantage* (numerically equal to the ratio in question) greater than unity. All machines function essentially in the same fashion. The saving of human exertion thereby can be very considerable, and herein lies the basis for a very large part of modern technology, for there is practically no modern piece of machinery, engine, or what not, involving moving parts, which does not somewhere incorporate in itself the principle of the simple machine.

The discoverers and users of machines must, however, have observed very early that the advantage provided by a machine does not occur unaccompanied by a compensating disadvantage, that nature is not inclined to give something for nothing. When a lever, for example, is used to raise a weight, the point at which the applied force acts must be moved through a greater distance than the weight is raised, in the proportion equal to the mechanical advantage. We may say that what is gained in the smaller amount of application is lost in the greater distance through which the application must be made. A similar situation prevails in the use of a pulley system like a block and tackle. To raise a weight with this by applying a force much less than the weight, the speed with which the pulley rope must be pulled is much greater than the speed with which the weight is raised, or alternatively, if one wishes to pull with a low speed, the time needed for raising the weight with the machine is correspondingly increased.

We mention this situation here because it proved to have an enormous significance in the future development of both technology and science, since it contains within it the germ of the concept of energy. This will be the theme of much of our next section. It is one of the first instances of the profound influence of technology on science, in this case, of course, physics through mechanics, its fundamental branch. Many others may be mentioned.

Think, for instance, of the technology associated with the reduction of metals from their ores. When one considers that this is a fairly complicated process usually involving the high temperature treatment of the ore with carbon in the form of charcoal, it remains a mystery how our primitive ancestors could have hit on it. It seems clear from the records that metallic copper was produced from its ores are early as 3000 B.C. Derry and Williams in their interesting history of technology[1] surmise that the first copper ore used was malachite, a green carbonate of copper. There is evidence that this was used as a pigment as early at 4500 B.C. If a little bit were dropped by accident in a strong wood fire, its easy reduction would produce a bead of metallic copper. This may have provided the clue which led to more systematic trials and hence to well-defined reduction methods. It was probably inevitable that the artisans who carried out this process would raise the question why it happened. There is no evidence that the question was answered for thousands of years. It will be recalled that the first clear-cut definition of a chemical element probably dates from Robert Boyle in 1660 A.D. The "elements" of the Greek philosophers (fire, earth, air, water) were quite another thing, though the concept was doubtless invented to account for the different appearance and behavior of common substances. In any case, the well-defined existence of technological processes, which we now call chemical, over centuries of time and longer, provided a pool of data to which scientific conceptions could be applied and tested. Other chemical illustrations are found in the use of dyestuffs, which in the first instance came from coloring matter from plants or animals; the production of soap from fats; the conversion of common salt into sodium bicarbonate, etc. In all these cases, our primitive ancestors must have been clever enough to realize that nature always exacts a *quid pro quo* for chemical processes as well as purely physical ones, that somehow the equivalent of human labor is involved in the heat usually required to make the reactions work. Once again, the significance of this for future science was enormous.

The use of drugs from plants in the treatment of disease was of course a technological activity of some importance. No one knows when people first began to collect herbs and dig roots in order to test their efficacy as remedies. But by the time of the great Greek

[1] T. K. Derry and T. I. Williams, *A Short History of Technology* (Oxford University Press, 1961), p. 115.

physician Hippocrates (born about 460 B.C.) the information about such medical materials was generally available and had accumulated through many centuries. The behavior of the drugs on the human system was as much a mystery as the action of machines, and it is not surprising that the kind of talk that prevailed about this was what we should now term a tissue of superstitions. Magic was invoked in the gathering of the material, and magic surrounded the processing and preparation. Yet the beginning of medical science was there. The same can be said for other medical techniques of a more pronouncedly physical character such as bleeding and purging. These also constituted early medical technology. Surgery goes far back in the experience of the human race as is evident from the presence of surgical instruments in early tombs.

The production of food was without question one of the most important objects of man's early technology. Only when people acquired the ability to produce food with such efficiency that it did not demand the whole of their waking time could civilization really develop. Anthropological studies indicate that this took place in the so-called neolithic age which set in about 3500 B.C., and was marked by the production of well-ground and polished tools and other implements of hard stone. Powers of observation must have been on the increase, since successful crop cultivation could only be assured by careful watch on the change of seasons and the weather. Hence, embryonic climatology and meteorology must have gone hand in hand with agriculture, though it was obviously shot through and through with magic and superstition. The need for an adequate water supply for crop growing was early realized, and crude devices for raising water out of a river and dumping it into an irrigation channel were devised and used before 100 B.C. Digging wells for water deep in the ground must have stimulated a certain amount of curiosity about the flow of liquids as well as about mechanical devices for raising water. Here again we see early technology providing the experience out of which science would ultimately develop.

We are accustomed to think of modern technology as involving the cooperative behavior of large groups of people, as, for example, in the assembly line of a manufacturing industry, and it is customary to contrast this to the essentially private character of primitive technology, in which a craftsman like a potter or a smith would make each piece of his product all by himself or at any rate with only a few assistants. The craftsman idea is of course a valid characteristic of early times, and many lament its apparent loss in big industry today. However, the difference is probably exaggerated. The mechanic who repairs your car or your refrigerator has to be a craftsman of sorts, even though he has more tools at his disposal than his predecessor of some thousands of years ago. On the other hand, cooperative technology was by no means unknown in antiquity, as the building of the Egyptian pyramids and temples amply attests. Still, it remains plausible that the advance of early technology was largely due to the independence of the individual craftsman, whose inventive urges were doubtless closely connected with his preoccupation with the problems of his trade and his desire to improve on his results. Progress probably came about as a sort of compromise between the conservative craftsman, who wanted to carry out the methods so carefully taught him by his master, and the discontented craftsman, who felt that by certain changes he could do better. It is unlikely that the "clever" people have changed too much in this matter since early human history.

We might go on and tabulate other specific examples of early technology, but this would be largely a recapitulation of history already available in many sources (for example, the book of Derry and Williams already mentioned). We are more interested in the relations between science and technology. Hence, in the following section we shall explore this, starting with the period in which something like modern science finally appears on the scene.

Mutual Relations of Science and Technology in Historical Perspective

It would be a fascinating exercise to trace from very early times the intricate interrelationship of science and technology. To do this would indeed entail a rather complete outline of the history of both disciplines. This is not the purpose of our book. We content ourselves with pointing out some significant trends with special reference to times not too greatly antedating our own.

As has already been intimated in the previous section, the situation in antiquity is not clear. The only thing we can say with some assurance is that technology emerged before science, simply because man had to overcome the problems of his environment before he had the leisure to indulge his curiosity about the world in an intellectual sense. The question then remains: what part did technology play in developing that interest in man's experience as a whole which we now call science? Here it seems the answer cannot be clear-cut. Certainly the behavior of tools, measuring instruments, etc., must have stimulated wonder why such things could be. Thus, the lever undoubtedly played a role in the establishment of the science of mechanics. We go into this presently. Similarly, the use of crude instruments for the measurement of land provided impetus for the development of geometry. On the other hand, the origin of fundamental ideas about the constitution of things, in which the Greek philosophers showed great interest, seems to have owed little or nothing to technology. Ideas like the "elements," e.g., the single element water as in the theory of Thales (ca. 600 B.C.), or the four-element (fire, water, earth and air) scheme of Empedocles (ca. 490 B.C.), later adopted by Aristotle, seem to have been pure feats of imagination. The same is true of the atomic hypothesis of Leucippus and Democritus (ca. 420 B.C.). It must be emphasized that these theories were not like the highly developed scientific theories of modern times. In particular, they led to no experimental tests and therefore made no demands on instrumentation, which might have provided a tie with technology.

Let us go back to the lever, since it provides an interesting illustration of the attempts of the Aristotelian school of philosophers to provide a theory of mechanics. The problem, of course, was to explain why a small force applied at a greater distance from the fulcrum leads to the exertion of a larger force at a smaller distance from the fulcrum. Now Aristotle and his followers had developed a simple theory for the effect of a force on a body, according to which the force is directly proportional to the velocity produced by the force. Symbolically

$$F = RV \tag{1}$$

where F is the force, V the produced velocity, and R a quantity representing the resistance to the motion: the greater the resistance, the smaller the velocity a given force will impart. Leaving aside the difficulties with this Aristotelian law of motion, we can see how it was applied in ingenious fashion to explain the behavior of the lever. The Aristotelians realized that when a lever moves about the fulcrum, the point at the greater distance from the fulcrum moves in the same time through a greater arc than the point at the smaller distance and hence moves with greater velocity. If now we may properly interpret what we have called the applied and exerted forces at the ends of the lever as resistances which must be overcome in the motion of the lever and treat F as constant in eq. (1), the smaller resistance is associated with the larger velocity and therefore operates farther from the fulcrum. Hence the well-known law of the lever emerges. Since eq. (1) is not a satisfactory hypothesis from the standpoint of modern mechanics, the deduction is illusory, even though it leads to the proper law. However, it is of some significance that the Aristotelians were aware that the velocities at the ends of the lever are directly proportional respectively to the lever arms (the technical term for the distance to the fulcrum) and that the product of each velocity by the force at the appropriate end remains the same for each end. Actually, this turns out to be a germ of the concept of energy, one of the most far-reaching concepts of physical science and indeed of all science. Whether the Aristotelians had any glimmering of the possible significance of this is unknown, but they must have realized that they were in the presence of an invariant, i.e., a quantity that stays constant in the midst of change. This was a notion much cherished by certain Greek philosophers, notably by Parmenides, as explained on page 155. At any rate, we have here a clear-cut illustration of scientific reasoning stimulated by and following on a purely technological device.

Aristotle and his followers were not the only Greeks who tried to give an explanation of the law of the lever. Archimedes of Syracuse, a much greater scientist, and indeed the greatest physicist of antiquity, also derived the law, though by a quite different method. Archimedes apparently had no faith that the reason could by itself suggest correct explanations for the *motions* of bodies, and he therefore limited his study to cases of equilibrium, i.e., bodies at rest under the action of forces, the branch of physics usually called statics. He felt that reason by itself dictates that if equal weights are suspended at equal distances along a lever from the fulcrum, equilibrium will ensue and neither weight will fall. The "reason" which is in question here is fundamentally the idea of symmetry. Since in this situation both sides of the lever are the same, there can be no reason why the situation should change, a kind of unconscious hypothesis, but a very powerful one, as modern physics has shown. His demonstration of the law of the lever based on this principle, so far as historians of science can understand it, runs somewhat as follows:

In Fig. 7.1 (a) we represent a lever AB equal in length to $3l$, with

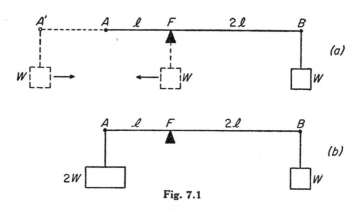

Fig. 7.1

the fulcrum F placed so that AF $= l$ and FB $= 2l$. A weight W is suspended from B. The question is: what weight must be placed at A to maintain equilibrium? Now it is clear to Archimedes that if we were to extend A to A′ where AA′ $= l$ and make the lever length $= 4l$ (recall the lever bar is assumed throughout to be weightless), we can maintain equilibrium by *symmetry* if we suspend W from A′ and another W from F. But then we can preserve symmetry and hence equilibrium if we move the W under F to the left a distance l and the W under A′ to the right a distance l, until they coalesce and form 2W under A. This yields Fig. 7.1 (b) and still provides equilibrium, since there is no loss of symmetry anywhere in the process. The demonstration is scarcely flawless and possibly not convincing to modern skeptics, but it at any rate provides another illustration of the way in which the challenge of a technological device was met scientifically.

The relations between science and technology during the long period between the classical age and the scientific revolution of the sixteenth and seventeenth centuries are far from clear. Most statements that can be made on the evidence available must be accepted with caution simply because it is clear that not all the evidence is in. Moreover, there are always difficulties of interpretation, as stressed in Chapter 5. However, it is probably not too wide of the mark to say that the fundamental science up to the time just mentioned was essentially that of the ancient Greeks, preserved largely by the Arab scholars and commented on by generations of scholarly monks in the monasteries of Western Europe. The Arabs and other Muslims themselves made some original contributions, notably by Alhazen in the study of the reflection and refraction of light. Technological devices and instruments were of course passed on and indeed in many cases without notable modifications for centuries. Thus, the clepsydra or water clock for the measurement of time, which dates from as early as 200 B.C., was still in common use in the tenth century A.D. As a matter of fact, Galileo used one in his experiments on the motion of falling bodies in the latter part of the sixteenth century. However, new developments in such instruments were gradually introduced. The invention of the mechanical clock dates approximately from A.D. 1000, and during the next three hundred years many were installed in cathedrals and other public buildings. Though the gearing of the wheels was reasonably expert and the energizing done by a falling weight, the control mechanism involving the so-called verge escapement was defective, and these clocks did not keep good time: some of them were out as much as two hours a day. Here we can cite a good example of the beneficial influence of science on technology. It was the scientific observation of Galileo (in 1581) of the isochronism of the small oscillations of the simple pendulum which ultimately led to the construction of reliable timekeepers, notably by Huygens.

On the other side of the ledger is the enormous help which technological devices offered to the new science of the sixteenth and seventeenth centuries. It is obvious that early chemistry could never have made headway without the possibility of precise weighing, and this was made possible by the availability of the balance, whose invention goes back to prehistoric times and which even as early as 1350 B.C. could weigh a gold coin of the order of ten grams with an accuracy of 1 per cent. The common suction pump, used for pumping water, dates back to Roman times. It was well known to Galileo and his successor Torricelli that this kind of pump will not raise water more than thirty-two feet. This peculiar behavior stimulated Torricelli to inquire into the reason. The ancients had indeed simply attributed it to nature's "horror of a vacuum." But Torricelli was not satisfied with this and was led to try the simple experiment of inverting a meter-long tube filled with mercury into a dish full of mercury, being careful not to allow any air to get into the tube in the process. Though the mercury was many times as dense as water, it did not all run out; rather, it continued to stand about 76 cm. in the tube, leaving a vacant space above it, now called the Torricellian vacuum. Torricelli found it hard to believe that any "horror" of a vacuum could hold up this mercury column. He thought it more reasonable to assume that the atmosphere pushed down on the mercury in the dish with just enough force per unit area (pressure) to counterpoise the thrust of the mercury column. In other words, he used the experiment to introduce a new concept, that of atmospheric pressure. This new concept was proved to be a very valuable one through the later suggestion of Pascal that if the instrument of Torricelli were carried up to the top of a mountain, the column of mercury would fall. When the experiment was tried, his prediction was verified, and the result of course was the invention of the barometer, a technological instrument which owes its origin to the new scientific thinking of the seventeenth century.

These examples could be multiplied many times. The microscope, the telescope, the thermometer, the air pump, the magnetic compass needle have all played important roles in the interaction of technology and science. These and many other cases show clearly that this interaction has not in general been one-sided but rather mutual. Without the instrumentation provided by technology the science of the sixteenth and seventeenth centuries could have made little progress, and indeed much of it grew out of the questions posed by these same instruments or related devices. At the same time in almost every case, the scientific discovery led to improved or new instrumentation and thus served to repay its debt to technology. This process proceeded at an accelerated rate during the period we have been considering and became so important during the late eighteenth and nineteenth centuries that we shall now pay special attention to what happened then.

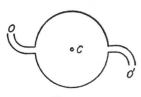

Fig. 7.2

To be specific, we first concentrate on the subject of heat and in particular its relation to work. On the technological side, the use of a hot object, e.g., steam, to do work or more crudely to make things move, goes back to the Greeks. Hero of Alexandria (ca. A.D. 50) showed that if water is heated in a vessel closed except for a small orifice, the flow of steam from the orifice could set into rotation a cylinder on the edge of which it impinged. Thus the idea of the steam turbine is a very old one. Indeed, Hero is supposed to have constructed a device of this kind (shown in Fig. 7.2) in which two necks with openings O and O' are attached to an otherwise closed vessel as in the figure. When the water in the vessel is heated and steam is produced, the flow out through O and O' produces a reaction which rotates the vessel counterclockwise. This is essentially what we should now call a reaction turbine. It is said to have been used to open temple doors. There is no record, however, that ideas of this kind were really exploited in antiquity.

The second Marquis of Worcester in his *A Century of the Names and Scantlings of Inventions by Me Already Practiced* (published in 1663) described a device for raising water by means of steam. The time was evidently ripe for this sort of thing. The Frenchman Denys Papin, who had been an assistant to Huygens, but who accepted in 1687 a professorship in Marburg in Germany, described in 1690 a steam engine in which a piston was moved in a cylinder both by the expansion of steam and the vacuum produced by its condensation. It is probably fair to consider it the first steam engine in the modern sense. In England, a short time later, Thomas Savery invented an engine something like that of Worcester for pumping water out of mines. His device was materially improved upon by Thomas Newcomen in the early years of the eighteenth century. James Watt and others added further improvements, until by 1800 steam engines had become fairly common pieces of machinery.

How much of this development was owing to the science of heat? All the available evidence indicates that it was very little. This point of view was expressed emphatically by a writer on the history of the invention of the steam engine, Robert Stuart Meikleham. In the preface to his book *Descriptive History of the Steam Engine,* of 1824, he wrote, "We know not who gave currency to the phrase of the invention being one of the noblest gifts that science ever made to mankind. The fact is that science, or scientific men, never had anything to do in the matter. Indeed there is no machine or mechanism in which the little that theorists have done is more useless. It arose, was improved and perfected by working mechanics—and by them only." Though the statement may sound exaggerated, candor compels one to admit that in the early days of steam engine development, the prevalent theory of heat was the caloric theory from which few really significant deductions about the properties of steam could be drawn. In particular, the latent heat of steam, that is, the amount of heat given up in the condensation of unit mass of steam to liquid water, was not discovered until about 1762, when the Scottish scientist Joseph Black made his epoch-making experiments on the thermal properties of substances. It is a fact that Watt was familiar with this work of Black and made some measurements of latent heat himself. One might then surmise that Watt used this knowledge in connection with his own improvements of the steam engine. It is reported, however, that he denied receiving any real help from the scientific work of Black and others of the time. We shall probably never know the whole truth about this matter. The practical mechanics who labored over the engineering of the steam engine naturally took pride in their achievements and disliked to admit that they owed anything to people who merely "mooned" over the constitution and properties of matter. The scientists on the other hand and the later historians of science just as naturally were tempted to exaggerate the influence of scientific concepts on technology. The situation is further confused by the fact that during much of the period we are considering and particularly in the nineteenth century the physical scientists were in general interested in technology and were by no means averse to making and discussing practical applications of their discoveries.

No matter what the complete truth of the dependence of the technology of the sixteenth to the nineteenth centuries on science may be, the fact remains that meditation on the practical problems posed by the steam engine initiated scientific developments of the most profound significance for modern physics, namely, in the theory of thermodynamics, already mentioned in a little detail in the previous chapter. It was eventually realized, though certainly not by the early developers of the steam engine, that the principal function of such a device is to transform heat into work. This was not an easy view to develop, simply because the nature of heat was for a long time the subject of uncertainty and controversy. Nevertheless, the early steam engineers did realize that one had to have a fire to make steam and the steam somehow managed to move something. They also realized that it took a lot of fuel to raise a given amount of water in one of the early water pumping engines. A very practical question then arose: to what extent can one increase the efficiency and economy of the process? It is likely that the early engineers felt that the matter was largely one of more careful mechanical construction, and certainly the efforts of men like Watt toward this end were rewarded by considerable success. This may well have stimulated the idea that by making tighter joints but at the same time reducing friction, etc., it would be possible to approach an ideal engine in which every bit of the heat which went into making the steam would be effective in the performance of work by the engine, a situation one might describe as 100 per cent efficiency. Now the thing we wish to point out, and this is of the utmost importance for a proper understanding of the relations between science and technology in the nineteenth century and our own time, is that the solution of this problem was not arrived at by the practicing experts in steam engines, but by a profoundly reflective scientist. He was the brilliant French physicist Sadi Carnot, who in 1824 brought out his memoir *Reflections on the Motive Power of Fire*. He showed by a logical process of thought akin to the demonstrations of the theorems of Euclidean geometry that the maximum efficiency of any heat engine does not depend at all on its mechanical construction or even on the working substance, e.g., steam, mercury vapor, or what not, but solely on the temperature at which the working substance enters the cylinder where it gives up its heat and the temperature at which it leaves the cylinder, either to be discharged into the surrounding medium or into a special condenser (as in Watt's famous invention). From this he concluded that there is no such thing as a 100 per cent efficient heat engine. To arrive at this result he had essentially to make use of the assumption that it is impossible for heat to flow continuously from a cold body to a warmer body without some form of other activity taking place in the environment. This was recognized later by Lord Kelvin, Rudolf Clausius and others as a form of the extraordinarily general principle called now the second law of thermodynamics. We discussed some of its ramifications in the previous chapter. Carnot's work was vital in the founding of thermodynamics. Of course, it had to be fitted into the work of many other men, like Rumford and Joule, Kelvin and Clausius, and made a part of the mechanical theory of heat, which looks upon heat as essentially nothing but a manifestation to our senses of the motion of the constituent parts of matter.

We do not propose to enter upon a historical account of the establishment of the theory of thermodynamics. What is important to stress is the fact that the desire for a better understanding of the behavior of a technological device, together with the wish to explore some purely speculative notions of a scientific character, can lead to the development of a wholly new branch of science. The illustration we have just discussed is by no means unique, but it happens to be one of overriding importance for twentieth-century science, since it served to introduce into science the concept of energy, probably the most general of all scientific concepts, for all natural phenomena are now regarded by scientists as transformations of energy from one form to another. At the same time, it has turned out to be the prime concept for technology as well, since the principal problem of technology is the energy supply. The social significance of this is enormous, as we shall see in the next section.

There is hardly space to cite more than a few other examples of the influence of science on technology during the nineteenth century. One is of special importance from the standpoint of both the science and the technology of the twentieth century. This is the discovery of electromagnetic induction by Michael Faraday in 1831. The Dane H. C. Oersted had shown in 1820 that surrounding a wire carrying an electric current there is a magnetic field, thus demonstrating a connection between electricity and magnetism which had long been sought. Faraday, a great experimental genius who was also a profound thinker, devoted his scientific career to the endeavor to detect relations between what appeared to be quite diverse physical properties. He reasoned that if an electric current can produce a magnetic field, it should be possible by proper manipulation of a magnetic field to produce an electric current. He succeeded in showing that the appropriate motion of a loop of wire in the neighborhood of a magnet does indeed lead to a current (electromagnetic induction). This suggested the large-scale production of electric currents, and generators were soon invented, based on the principle Faraday had established. It seems highly unlikely that such a device would have been invented if it had not been for the prior scientific investigation of a scientist like Faraday, who, as it happened, was not interested in any practical application but wished solely to enlarge the realm of human experience. His results showed, however, the practicality of the transformation of mechanical energy into electrical energy and led to the foundation of an entirely new branch of technology, namely, electrical engineering. Fifty years after his discovery central power plants were beginning to supply electrical energy to large cities for lighting and ultimately other purposes.

At the same time, Faraday's discovery and others in the same realm gave great impetus to further scientific investigations along similar lines. Clerk Maxwell translated Faraday's ideas into more abstract and powerful mathematical language and, by adding some ingenious notions of his own, produced the theory of electromagnetism, which in turn led directly to the electromagnetic theory of light, the view which considers light to be essentially propagated as electromagnetic radiation. As everyone knows, the working out and experimental verification of this theory was responsible for the creation of radio broadcasting as a branch of technology. Though no one doubts the significance of this in twentieth-century society, a more important result of Maxwell's theoretical invention was the Einstein theory of relativity, with its momentous consequences for both the science and technology of our time.

To come back to thermodynamics, we can mention one final illustration of the impact of science on technology, and this will have to suffice for the present section. In the hands of Willard Gibbs, thermodynamics became a powerful tool for the study of chemical reactions. His famous memoir *On the Equilibrium of Heterogeneous Substances* (1875-1878) formed a large part of the basis of physical chemistry and thus led to the establishment on a firm scientific basis of what has come to be called chemical engineering.

Industrial Research in the Twentieth Century

In the light of the previous section, it would be natural to suppose that technological developments which emerged from the scientific discoveries of the early and mid-nineteenth century would be undertaken primarily by the scientists themselves or at any rate guided by them in close collaboration with engineers. However, with a few striking exceptions this was not the case. Faraday did not follow up his celebrated discovery with the exploitation of the electric generator, nor did Maxwell engage in any practical application of the electromagnetic waves whose existence he predicted. Even Hertz, who verified Maxwell's prediction by experimentally producing electromagnetic waves of lower frequency than that of light and showing that they have all the properties of light save the ability to stimulate the sense of sight, did not try to exploit his discovery industrially. The same sort of thing goes for Gibbs. In fact, during his lifetime very few people understood his researches sufficiently to grasp their possible practical application. It was men like Edison in the United States, Ferranti in Great Britain, and W. von Siemens in Germany, all inventors and engineers, who put electrical engineering on its feet and devised methods for the large-scale generation of electrical energy. It was Edison again and Swan in England who invented a satisfactory incandescent electric light to be powered by the distributed electrical energy. It was people like S. F. B. Morse in the United States and W. F. Cooke in England who invented the electric telegraph. The first practical use of electromagnetic waves for the transmission of information through space was made by the Italian electrical engineer G. Marconi. The telephone we owe to, among others, Alexander Graham Bell. All these people were primarily inventors and not scientists in the sense which science had come to mean in the nineteenth century.

There were of course exceptions to this general rule. One thinks at once of Lord Kelvin, one of the great physicists of the nineteenth century, who did not hesitate to turn his grasp of the scientific principles of electricity and magnetism to good account in the design and manufacture of electrical measuring instruments as well as an improved magnetic compass. His contributions to the solution of the problems involved in the Atlantic cable are also well known. He was unusual among physicists of his time in that he made money out of his inventions stemming directly from his scientific work. Another possible exception to the general trend we have mentioned was Sir Charles Wheatstone, who was both a teacher of physics and a practical inventor in the field of telegraphy. However, in his case, his purely scientific achievements were overshadowed by his technical contributions.

In general, it is safe to say that the nineteenth-century scientist, like his predecessors of earlier ages, even when he turned up phenomena which seemed to have exciting possibilities for technological exploitation, preferred to leave this phase alone and to stick to the description, creation and understanding of experience. The famous American physicist Joseph Henry, independent discoverer of electromagnetic induction, is a good illustration of this tendency. It was largely left to other clever people who had grasped enough of the scientific results to understand their significance to undertake to apply them to some kind of gadgetry with possible social value. It is indeed striking that often such persons in the course of their inventive activity ran across new experience of utmost scientific value but passed it by. Thus, Edison in experimenting with the incandescent lamp undoubtedly discovered thermionic emission, that is, the emission of electrons from heated metals. But he failed to follow this up as a scientific phenomenon. This is a bit ironic since the effect has served as a basic for the thermionic vacuum tube, so useful in the radio industry.

We must mention a fundamental contribution of Edison which was possibly of as great ultimate social significance as his ingenious inventions. This was the establishment of a research laboratory in Menlo Park, New Jersey, in 1876. Here, for the first time in history, so far as can be ascertained, he set up an organization for the specific purpose of systematic invention. This was the forerunner of the modern industrial research laboratory, which has revolutionized the relations between science and technology in the twentieth century. It should be made clear that Edison's laboratory was not a laboratory for scientific research. His sole purpose was to dream up and then produce gadgets which would have economic value, i.e., could be sold at a profit to a public which found them useful or exciting. So far as can be seen, he had no interest in studying experience for its own sake. He wanted practical people on his staff and was rather distrustful of "long-haired" scientists. Nevertheless, his laboratory had a well-defined program and pursued it systematically. It made possible the much more efficient exploitation of inventions than could ever be achieved by the lone inventor working in his cellar.

It was inevitable that with the increasing potentialities of electricity for public exploitation as the nineteenth century moved on to its end, there would arise in the United States large companies to provide the materials. Thus, in 1892 the General Electric Company was formed as a merger of two smaller companies, the Edison General Electric Company and the Thomson-Houston Company. The former devoted itself to promoting the incandescent lighting system developed by Edison, while the latter manufactured electric arc lights and the associated systems as developed by Elihu Thomson, the inventor and applied scientist, who obtained more than seven hundred patents for his inventions. It was soon recognized by the directors of this new company that the amount of technological development which could be drawn out of the scientific knowledge already accumulated during the nineteenth century, though large, was finite and that there would be a greater chance of ingenious developments if there were more science to work with. But this meant that they needed more basic scientific research, essentially the kind of research which one would expect to find done in university laboratories by scientists anxious to find out how things go just for the joy of knowing. One might naturally raise the question, "Why did not the General Electric Company engineers study the research output of the American university science laboratories and endeavor to find something there they could use?" The answer is that by and large the university scientists were not doing very noteworthy research in this country at that time, or if they were, it did not seem relevant to the needs of the General Electric Company. That scientific research in physics, for example, was indeed going on at this time in the United States is evident from the work of men like Gibbs, Rowland and Michelson. But no one in this country could see what Gibbs was driving at; the appreciation of this really came first from abroad. Rowland and Michelson were interested in precision measurements in optics; they developed some marvelous instruments, i.e., Rowland's engine for ruling gratings for the production of optical spectra and Michelson's interferometer for precise experiments in light. But these things were not precisely what the General Electric Company was looking for. They really wanted someone who was interested, among other things, in the scientific problems involved in making a better incandescent lamp. It is clear that they wished to approach this from a different standpoint from that of Edison, which boiled down to trying in patient painstaking fashion every conceivable scheme for solving a given problem until one hit on the best. There was evidently a feeling in the air that the same techniques which were leading to the successful progress of pure science could be successfully applied to practical problems, so as greatly to economize laboratory effort on the one hand, and by the pursuit of uncommitted inquiry, so characteristic of science, strike out in entirely new directions, from which something useful was sure to emerge eventually.

So the General Electric Company decided to set up its own research laboratory in the year 1900. They were extremely fortunate in persuading a young physical chemist, Dr. Willis R. Whitney of the Massachusetts Institute of Technology, to head the new organization in Schenectady, New York. He rapidly gathered around him a group of clever scientists who had the professional background and insight to glimpse the essential scientific character of practical problems in technology and by being encouraged to follow their own scientific hunches eventually made discoveries of the greatest industrial importance. The story of Irving Langmuir's study of the effect of various gases on the behavior of incandescent lamps is typical of what happened at the Research Laboratory of the General Electric Company. The whole story is too long for the telling here, but the essential point is that Langmuir spent several years studying the effect of gases in the lamp bulb on the radiation of thermal energy from the light filament as well as the rate of evaporation of the filament material. In this work he turned up some very interesting results concerning the conduction of heat through gases and other purely physical phenomena which might to a less ingenious person seem to have little practical relation to the more efficient operation of a lamp. But the fact of the matter is that these purely scientific investigations paid off in the form of the production of the much improved Mazda lamp. Other illustrations of the value of industrial research in the experience of the General Electric Company[2] include the investigations of W. D. Coolidge on methods for making the metal tungsten ductile to improve its availability for use in the incandescent lamp; the researches of the same physicist on the use of the hot tungsten cathode to provide greater control in the output of X-ray tubes, which revolutionized this important source of radiation in medicine; researches on the emission of light from hot sodium vapor; studies on the emission of electrons by hot metals (thermionic emission), in which it was discovered that the addition of a small amount of thorium to the tungsten filament in a radio tube could increase its efficiency many times; the investigations of Langmuir on the physical and chemical properties of surface films (many only of thickness equal to the diameter of one molecule, i.e., monomolecular layers), which led to the development of the whole new field of surface chemistry and won for Langmuir the Nobel Prize in 1932. The list could be extended considerably, coming down to the work in nuclear energy dating from the Second World War and since. It would include research in virtually every major field of physics and chemistry. The significance of all this from the standpoint of the relations between science and technology is twofold. In the first place, all the researches carried out in the Laboratory (save those directly connected with military applications, governed by security regulations) were reported to the scientific community through papers presented at meetings of scientific societies and printed in the scientific journals. This made a material addition to the sum total of basic scientific research in the country, which had hitherto been confined largely to the universities. In the second place, almost all the investigations ultimately proved of technological value, some very quickly indeed and others after a lapse of time. This obviously justified the expense of the maintenance of the Laboratory in the eyes of the stockholders of the General Electric Company and stimulated other companies to establish similar research laboratories.

The first quarter of the twentieth century saw the establishment of

[2] For details see Laurence A. Hawkins, *Adventure into the Unknown—The First Fifty Years of the General Electric Research Laboratory* (New York: Morrow, 1950).

most of the great industrial research laboratories of today, including, in addition to that of the General Electric Company we have just been discussing, that of the DuPont Company in Wilmington, Delaware; the General Motors Laboratory in Detroit, Michigan, in 1909; the Bell Telephone Laboratories in New York in 1910. Other large organizations established during the same period included the Westinghouse Research Laboratories in Pittsburgh, Pennsylvania, and the Eastman Kodak Laboratory in Rochester, New York. According to C. E. K. Mees[3] there were about 300 industrial research laboratories, both large and small, by 1920. The number had increased to over 2200 by 1940. In the years since World War II there has been further expansion. In a study carried out by the National Science Foundation[4] and published in 1960 as one of its series *Surveys of Science Resources*, 6800 companies in the United States were studied with respect to the funds they devoted to research and development. The resulting figures are of great significance in connection with the progress of industrial research. Before citing them, we must make clear the meaning of the words "research and development" in this context. For the purposes of its report the National Science Foundation divides research into two categories as follows:

Basic Research. This includes "original investigations for the advancement of scientific knowledge that do not have specific commercial objectives, although such investigations may be in fields of present or potential interest to the reporting company."

Applied Research. This includes "investigations directed to the discovery of new scientific knowledge that have specific commercial objectives with respect to products or processes."

On the other hand, the Foundation defines "development" to mean:

Development. This includes "technical activities of a nonroutine nature concerned with translating research findings or other scientific knowledge into products or processes. This does not include routine technical services to customers or routine product testing."

It is not difficult to give illustrations of these categories. The previously mentioned researches of Faraday on electromagnetic induction

---

[3] C. E. K. Mees, "The Path of Science," (New York: John Wiley and Sons, 1946).
[4] *Funds for Research and Development in Industry, 1957*, NSF Report, 60-49.

and of Maxwell and Hertz on the nature of electromagnetic radiation were certainly examples of basic research. The work of Langmuir, on the other hand, on the behavior of an electrically heated filament in an evacuated glass bulb was applied research. The construction of a prototype lamp based on the results of Langmuir's research was development. This does not of course extend to the large-scale manufacture of such lamps once the prototype has been designed, made and tested to see whether it behaves as predicted.

It seems clear that most industrial laboratories will be concerned with applied research and development. However, this is not wholly the case, as was suggested by our brief review of the General Electric Research Laboratory and as will further become clear in a moment. In any case, let us now look at some figures. In the National Science Foundation report just referred to, total funds expended for research and development in the industries of the United States came to 3.4 billions of dollars in 1953. In 1957 this figure had risen to 7.2 billions. These are substantial sums, even if they are very small fractions of the so-called gross national product. It is interesting to note that of the 1957 sum the Federal Government contributed slightly more than half, mainly through contracts for national security purposes (e.g., aircraft and related things). It is also interesting to observe that of the total amount for 1957, basic research accounted for about 250 millions of dollars, applied research for 1.5 billions, and development 5.5 billions. The NSF considers that industrial laboratories *do* carry out basic research, although the total support of this type is rated at a small fraction of the amount expended for research and development as a whole. Another point brought out by the NSF report, not without social significance, is the fact that over 6 billions of dollars of the total 7.2 billions expended in 1957 (some 85 per cent) were expended by companies with total employees numbering 5000 or more and 90 per cent of the total was accounted for by companies employing more than 2500 people (not just in their laboratories, of course, but in all phases of their work). Though a discussion of the precise distribution of the research and development funds would take us too far afield here, it is worthwhile pointing out that the aircraft, electrical equipment, motor vehicle and other transportation equipment and machinery industries account for some 72 per cent of the total funds for 1957. The chemical industry, including medical drugs, accounted for only 8 per cent, and professional and scientific instruments, for only 4 per cent.

These figures provide food for reflection. First, there is the very large increase from 1953 to 1957 with the implication of an accelerating rate of progress: technology would appear to be finding its association with science fruitful. On the other hand, the relatively small fraction of the total expenditure devoted to basic research is not such a favorable omen. Even more questionable is the fact that the increase in basic research support by industry did not increase so fast in the interval 1953-1957 as did the overall research and development expenditure. Let us look into this matter a little.

It was natural for the early industrial research laboratories to concentrate largely on applied research, since they had definite problems to solve and needed the scientific answers for these. These were adequately brought out in our description above of the activities of the General Electric Research Laboratory. However, as time went on, it was realized that really *new* developments can come only from essentially *new* ideas. But these as a rule have emerged only from the same type of uncommitted research as that normally carried on nowadays in university laboratories, where the investigator chooses a field of research because of his interest and not because he thinks his work will solve a practical problem. The Bell Telephone Laboratories were among the first to translate this realization into practice. Thus, instead of merely trying to improve the performance of telephones from an engineering point of view, they early began to concentrate attention on fundamental studies of speech and hearing. This led to previously unsuspected relations between the sensitivity of the ear and frequency, for example (discussed on page 139). It also turned up new knowledge on the acoustical character of speech sounds. All this was basic, though its practical application to communication was not long in following. Another field of basic study at the BTL was magnetism, in many ways a neglected subject in university research. Here, entirely new conceptions were worked out, like the concept of magnetic domain, and here again the development of new magnetic alloys, like permalloy, with remarkable properties soon followed. It was studies by BTL physicists, notably C. J. Davisson and L. H. Germer, on the reflection of electrons from the surfaces of single metal crystals, which indicated for the first time that electrons, though normally thought of as particles, can be diffracted by crystal lattices like waves. This piece of basic research was one of the experimental foundations of the theory of quantum mechanics, the basic atomic theory of the twentieth century.

After the mid 1920s the claim of basic research to be an integral part of the program of every industrial research laboratory was uncontested, and there are few laboratories in the United States which do not employ some scientists for this purpose. This has "paid off" in the economic sense that the time gap between the attainment of basic results and their successful application in development of new products and processes has steadily shrunk. Recall that it was a whole half century after Faraday's discovery of electromagnetic induction (1831) that Edison opened his Pearl Street electrical generating station in New York (1882). Contrast this to the case of nuclear energy: nuclear fission was discovered in 1939. The first so-called atomic bomb was exploded in 1945. The development of the transistor as a new type of electrical valve (it threatens to displace the thermionic vacuum tube) has followed even more rapidly on the basic research on the properties of semiconductors. There is therefore no exaggeration in referring to the vanishing frontier between basic research on the one hand and applied research and development on the other.

But the careful observer will raise a question: if basic research is so important in our time, why is it that of the total industrial research and development expenditure in 1957 less than 5 per cent went for basic research? If we exclude development and consider research alone, only about 15 per cent of the total research outlay was for the basic variety. How can one account for this? There is, to be sure, an easy answer, and that is to say with concerned annoyance that industry in general has not learned the lesson of the twentieth-century alliance between science and technology and is foolishly neglecting its opportunities. There is some justification for this in certain quarters, but before we get too excited over it, we should consider the simple fact that in practically all fields basic research is less expensive than applied research and development. It may take only one superior scientist to get a bright idea and with a few assistants to test it experimentally or to compare its consequences with experimental results already in the scientific literature. It can well demand twenty-five to fifty people of varying grades of technical competence to work out the details of the idea so as to relate it to possible practical consequences. Finally, it probably will take several hundred people to develop the practical result of the idea into a marketable device or process. The differential cost factor then no longer seems so serious. Moreover, equipment cost for basic research may be expected to be far less than in applied research and development. To be sure, one should not be completely satisfied with this answer. It neglects, for one thing, the problem of the proper distribution between the two research activities in any particular industry. It may well be that there are cases where very imaginative industrial scientists are being forced to work on applied research and to direct development, when in the long run they would contribute more heavily to technological development by devoting full attention to basic research. The matter tends to be complicated also by the fact the salaries tend to be higher for research *direction*, which obviously goes more appropriately with applied research. It is not at all easy to direct fundamental, uncommitted research. One can discuss it with the investigator, one can give advice and sympathy, but on net balance the really productive scientific thinker has to be a lone wolf.

Our discussion of industrial research in this section has been confined to American industry. This phenomenon is, however, by no means unique to the United States. The same sort of thing has happened in Europe and Japan, though possibly on a smaller scale and a little later in time. Perhaps the best example in Europe is provided by Philips Research Laboratories in Eindhoven, Netherlands. The Philips company began with the lamp-making business. When the First World War cut off their supply of glass from Germany, they began research on glass, and this ultimately led to the foundation of their research laboratory in 1923. In 1914 the firm had four scientific researchers.[5] This number grew to 165 in 1931 and to 415 by 1936. Since the end of World War II the laboratory has been directed by one of the most distinguished theoretical physicists in the Netherlands. It has enormously broadened its research scope and has established subsidiary laboratories in many parts of the world. Its scientists publish their results in all the leading physical and chemical journals in the world.

[5] See J. G. Crowther, *The Social Relations of Science* (New York: Macmillan Co., 1941).

We may summarize the essential content of this section by repeating that on our contemporary scene the mutual influence of science and technology has grown to be decisive and immediate. The new ideas resulting from basic scientific research become translated into practical devices within a fantastically short time. At the same time we must not overlook the very important assistance which the new technology renders to science. Thus, for example, the scope of experience which is now accessible to scientific investigation has been enormously increased by technological methods for producing very high and very low temperatures as well as similar extremes of pressure. The electrical industry now manufactures high-voltage machines capable of accelerating charged atomic particles to velocities approximating that of light and hence with relatively enormous energies which can produce a host of nuclear disintegrations and provide a vast amount of experimental evidence in the subnuclear realm for theorists to try out their ideas on. So technology continues to repay its debt to science, and the relations between them become ever closer.

The Role of Technology in Civilization

From the very beginning in Chapter 1, we have stressed that science makes its impact on society in two ways. One of these is ideological and the other technological. Since most people fail to grasp the first but see the second without trouble, we have devoted a good deal of the argument of this book to the ideological or philosophical role of science in society. But we obviously must not overlook the second method, and our discussion in the present chapter on the relations between science and technology clearly sets the stage for a more careful examination of just how technology affects society.

If we treat the matter in terms of devices and processes, things become very complicated indeed. It is far more satisfactory to examine what technology has done and is likely to do in terms of the concept of energy, which we have already mentioned as the most significant of all the concepts of science. From this point of view, what the growth of technology has done is to increase the number of different transformations of energy available to man and to make available more energy for the transforming process.

To primitive man the principal energy transformations at his disposal were those connected with his own body and those of his neighbors, including the lower animals, with the addition of course of biological transformations involved in the growth of plants brought about by the energy of solar radiation. He was of course affected by those transformations, due to the same source, in which water changes its phases from liquid to solid or vapor. As soon as he discovered fire, primitive man used another type of energy transformation, namely, that involved in what came to be called chemical reactions, the change of one substance into another. The domestication of animals gave him an energy supply independent of his own efforts, and he put this to use in the transformation from rest to motion implied in transport, building and the manufacture of domestic articles. Later still, the discovery of the power latent in flowing water, in the wind and in fuel which he could dig from the ground, and whose energy he could transform into motion by means of a heat engine, gave man a vast increase in the number of what some writers have called his "energy slaves."[6] For example, Ubbelohde quotes from another source the increase in the annual energy consumption from coal per head of population in Britain from 1800, when it was the equivalent of one ton of coal per year, to 1952, when the corresponding figure was about five. In the past two centuries indeed the growth in the energy supply in the so-called industrialized countries has been phenomenal and may be taken to symbolize the advance in material civilization due to technology.

It is this vast increase in the energy supply which has made urban life possible for huge populations and more recently has facilitated its extension to suburbia. At the heart of these urban complexes stand the electrical power station, the water pumping and sewage disposal plants, the gas pipe line and the gasoline and diesel oil stations. All these place at our disposal numberless energy slaves ready to do our bidding at the turn of a cock or the flick of a switch. Supplementary to these large scale energy supplies are those involved in communication, viz., the telephone, radio and television. These too use up energy, albeit at a smaller rate per head of population. Civilized man in the twentieth century has got himself into a mental

[6] See, for example, A. R. Ubbelohde, *Man and Energy* (London: Hutchinson, 1954).

state where he takes all these energy slaves for granted, and when for any reason some temporarily cease to serve him, his dejection becomes profound. He apparently rarely realizes how lucky he is to have all this, which can be bought in Western civilization with relative ease, and often is tempted to listen to political demagogues who strive to convince him that it all costs too much, in spite of the fact that it has brought him leisure to develop his own interests undisturbed by the crushing load of labor which was the lot of the vast majority of his predecessors not many years ago. Though one would suppose that contemporary civilized man should consider himself lucky with all his energy slaves, a simple glance around indicates that all this has not proved to be an unmixed blessing. More energy has meant more problems of all sorts. We expect to discuss these in summary fashion later in this section. But first we look into a practical problem. Assuming that it is the vast increase in the energy supply which is really at the basis of modern material progress, what assurance do we have that the supply is sufficient to meet our needs, real and fancied? This demands a little examination of the nature of the supply.

The first point to make clear is that the source of all energy available on the earth is the sun. This goes not only for the current supply in the form of solar radiation reaching the earth's surface but also for the various forms of capital resources stored in the earth or its atmosphere. It is estimated that the sun is radiating energy at the enormous rate of about 4×10^{23} kilowatts. Recall from the previous chapter that the watt is the fundamental unit of power used in scientific work and is equivalent to the rate of energy expenditure of 1 joule per second. The kilowatt is of course 1000 watts. The joule is the large-scale unit of energy equivalent to 10^7 ergs. The significance of the power output of solar radiation can be appreciated to some small degree by remembering what a 100-watt electric bulb is like. Of course, only a very small fraction of the total solar radiation reaches the earth and penetrates the earth's atmosphere. The average rate of flow of solar radiation reaching the earth's surface is estimated to be about 10^{14} kilowatts. In 1950 it was estimated that the total electrical energy available per head of population in the United States for the whole year was about 1870 kilowatt hours. (Note that the kilowatt hour is a unit of energy equal to that supplied in one hour by an agent working at the average rate of 1 kilowatt during that time. It is equal numerically to 3.6×10^6 joules.) This electrical energy supply is equivalent to an average power allotment per head of population of about 0.2 kilowatt. The United States ranks high in the availability of electrical energy to its population, though on a per capita basis it is exceeded by Norway, Sweden and Switzerland, for obvious reasons. In any case, the contrast between 0.2 and 10^{14} kilowatt is striking, to say the least. If only we could use even a small part of this steady stream of solar energy which pours in on the earth's surface! As a matter of fact, of course, we do take advantage of it, since by the process known as photosynthesis it provides for all our agriculture. We also use it, though perhaps apparently less directly, when we operate windmills. Water power is another example of the availability of solar radiation for current energy supply. But even these are very small fractions of the total available. Thus it is estimated that the potential water power for hydroelectric plants is about 2×10^8 kilowatts, or about half a million times smaller than the total solar power available. Of this, perhaps only slightly more than one tenth has actually been tapped. Various attempts have been made to employ direct solar radiation for cooking food and heating homes. So far success has not been too great, since meteorological fluctuations prevent continuous performance. Efficient means of storing solar energy during the periods when the sun shines and releasing it during other periods are badly needed if we are to take advantage of what amounts effectively to an inexhaustible energy supply.

Actually, at the present moment man is depending principally on capital energy reserves in the form of fossil fuels, i.e., coal, oil and natural gas. These represent energy stored in the earth through the effects of solar radiation over long periods of past time. The energy so stored can be made available only by mining the coal from the inside of the earth and drilling wells for and pumping the oil and gas. The energy is finally released by the process of combustion, which involves the chemical reaction known as oxidation, i.e., the rapid combination of the fuel with the oxygen of the air. Technologically it is wasteful and not very efficient, but it is what the people of this planet are now principally relying on to provide their energy slaves. In the year 1957 the United States alone burned about 470 million metric tons of coal and lignite (the metric ton is 1000 kilograms, which is about 2200 pounds or somewhat greater than the English ton); it also burned crude oil, gasoline and natural gas to the *equivalent* of 1040 million metric tons of coal or the equivalent of 1.7×10^{12} kw hours of energy. In 1950 it was estimated that the proved oil reserves of the world, that is, the total oil believed to be recoverable under existing economic conditions, was about 4.75×10^{12} gallons with a total estimated energy content of about 2.8×10^{14} kw hours. Unfortunately, these figures are unrealistic, since they take no account of new and active prospecting for oil, a very vigorous business. Previous predictions that the world would run out of oil in a relatively short time have been negated by the activities of oil companies. Nevertheless, the fact remains that no matter how much oil remains to be discovered (and some of it will doubtless be in places difficult to work) the total supply on the earth economically available to man is finite. When one lives off capital, it eventually disappears. The same situation prevails with coal, though here the time scale is admitted to be much larger. The energy in the world's proved coal reserves is estimated to be about thirty-four times that in the oil reserves. The tendency in highly industrialized countries like the United States is to turn to oil instead of coal as being more convenient and possessing more energy per unit mass. The ultimate distillation of coal to provide gas may provide a solution to the coal problem. Fossil fuels will be around and will be used for a long time to come, but by their very nature their ultimate exhaustion is inevitable and their replacement essential.

A more promising source of capital energy supply is that trapped in fissionable nuclear material. The discovery in 1939 of nuclear fission, with the relatively large energy release per split nucleus, at once aroused hopes that the world's energy supply problem would find a solution here. To appreciate the significance of this, we must recall that a chemical reaction like the oxidation of carbon in the burning of coal involves *per atom* involved an energy release of only a few so-called electron-volts, where the electron-volt is equal to 1.6×10^{-19} joule in the conventional units above or about 4.5×10^{-26} kw hours. This seems small and it is, though we must remember that the figure relates to the energy output per atom involved in the reaction. If a gram atom of material is involved, i.e., a mass of the reacting substance in grams equal to the atomic weight (which for carbon is equal to 12), we must multiply this figure by the number of atoms in a gram atom, which is the same for all elements and equal to the so-called Avogadro number, namely, 6.02×10^{23}. Now the striking thing is that in the fission of a uranium atom by a neutron the energy released per fission can be as much as 200 million electron volts. Evidently, this process if it can be carried out on the same scale as ordinary chemical combustion is a vastly superior energy producer, weight for weight. As is well known this is the process used in the first atomic bombs. At the same time, it is also being employed in so-called nuclear reactors for the production of power in power plants. Some half dozen of these power reactors have been built in the United States and Western Europe using uranium as the "fuel." The new reactor-powered submarines come to mind as illustrations. At the moment, such reactors cannot compete economically with power plants energized by fossil fuels, but the future in this respect looks hopeful. Uranium is very plentiful in the earth's crust, and this assures an energy supply which, though also capital in character, can long outlast oil and probably coal as well.

Much energetic experimentation is now taking place in the effort to apply the so-called thermonuclear or fusion process to the production of energy on a large scale. This in brief is the reaction by which hydrogen nuclei are joined ("fused") to make helium with the accompanying emission of a large amount of energy, viz., about 28 million electron volts per helium atom produced. In view of the relative lightness of the helium atom, this form of energy release is much more powerful than uranium fission, weight for weight. Since the supply of hydrogen on the earth (from sea water) is virtually inexhaustible, if the thermonuclear reaction can be realized it will solve the problem of the energy supply on earth as long as the planet remains habitable by man. Unfortunately, the process, as the name suggests, demands very high temperatures for its carrying out (i.e., of the order of millions of degrees centigrade), and so far this has been possible to realize only in the hydrogen bomb, which is hardly sufficiently tame for use in a power plant. At the moment, the solution of the engineering problem involved seems far in the future, but there is no reason to suppose that it will not ultimately be attained.

The conclusion we arrive at is that science coupled with technology can solve the problem of the energy supply for the foreseeable future and hence that material progress is assured, no matter how imaginative and enterprising governments and private entrepreneurs in the times ahead may be. Does this mean then that technology will solve all our problems and we shall all live "happily ever after"? Unfortunately, no. Experience shows that though cheap and plentiful energy solves some problems, it also introduces problems just as crucial. Cheap energy has meant a decided increase in the food supply; but we are witnessing the pressure of increased population on this increased food supply which Malthus emphasized so many years ago. Admittedly, there are mysteries here which the invoking of any particular sociological principle is inadequate to account for, but there seems to be no doubt that we are faced at this time with a population explosion of unprecedented magnitude. It is probable that the energy supply can be increased rapidly enough to feed these people enough to keep them alive, if not well nourished, just as modern medical technology is going to keep a larger percentage of them alive to maturity by improved sanitation and public health procedures. But the more serious problem is: what is the world going to do with all these people, or put in another way, what can they find to do to justify their existence? One of the prime characteristics of advancing technology in all ages has been the release of human beings from back-breaking labor. This is now accelerating to the point where in many business enterprises human labor is becoming almost unnecessary, except in a supervisory capacity. The displacement of human beings by machines, known as automation or more accurately automatization, is a particularly fast-moving technological development of modern times. It represents the attempt to replace human control of energy supply by automatic control with consequent gain in speed and accuracy. Though the introduction of such methods may well lead to jobs for more skilled individuals to produce and service the corresponding equipment, there seems to be no question that their first effect is to throw out of work numbers of unskilled people for whom further jobs appropriate to their abilities may be hard to find. It should be emphasized that automation involves not merely the introduction of automatic control machinery to insure a "manless" factory but a complete rethinking of the whole process of operating a given industry so as to take advantage of every possible technological process which can lead more efficiently and faster to a better and more salable product. This of course involves not only manufacturing operations but also those associated with cost accounting, sales, and so on. Strictly speaking, a completely automated industry would throw out of employment not only factory personnel but also administrators save for the very highest echelons. They alone with maintenance and repair crews would run the business. Needless to say, this state of affairs is still in the discussion stage! But it will almost certainly come.

Almost every energetic businessman, not to mention the social scientist, can present pet solutions to the difficulties involved in automatization and ultimate automation. We do not propose to add to the number but are content merely to point it out as a concomitant of technological advance. Closely associated with it is the problem of the "leisure" time made available by the vastly increased productivity. It has been assumed by some thinkers that people would more or less automatically know what to do with the leisure thus gained. But this has not turned out quite so simply. No one can say what the right use of leisure is, but it certainly seems as if many persons use it merely to mull over their real or fancied troubles with accompanying jeopardy of their mental stability. It is true that vast new industries of entertainment have sprung up hand in hand with technological advance, but the massive demands made on them by the increasing leisure of the masses seem to be leading to diminishing returns of everything save utter boredom. And boredom plays directly into the hands of the psychiatrist.

There are many in the camp of the humanists whose most damning indictment of modern technology (which, as we have indicated, they often confuse with science) is the role they think it plays in forcing conformity in the social habits of people. The mass media, skilfully directed by Madison Avenue and its conscientious imitators throughout the civilized world, dish out every day the same urgent "gospel of salvation through stuff" to millions of folk, who, though many may scoff at it, nevertheless in the main tend to accept it and behave accordingly. Those who consider their tastes refined deem the message banal and vulgar and the results utterly deplorable. Technology gets the blame. One can sympathize with this point of view and still feel that the emphasis is misplaced. Grant that technology has placed at our disposal enormously increased facilities for communication. If man abuses these as he has abused every means of communication developed in the past, should technology be held responsible? The argument seems to reduce to a mere wringing of hands with a little gnashing of teeth thrown in for good measure. (See the discussion of "Literature," page 62.)

Another genuine fear of those who have doubts about the beneficial influence of technology or society is that it will inevitably lead to more state control and a rigidly planned economy. Now it is undoubtedly true that whenever you make generally available complicated gadgets, some measure of social control is necessary to see that not too many people hurt themselves or others. There is nothing essentially new about this. However, the more numerous and more complicated the gadgets, the more elaborate the controls become. When the horse and buggy were the principal mode of conveyance about town, traffic control could be relatively simple or even nonexistent. In the automotive era it is necessarily more detailed. On the whole, people have submitted to this more patiently than might have been expected. It is true that thousands of them manage to kill themselves and others and still more thousands get injured on the roads of the world every year. This is lamentable, and every effort by reasonable control should be made to cut the number down. But a judicious statistical examination of the figures in the light of the enormous increase in traffic should arouse not so much horror that the damage is so great but rather wonder that it is no greater. Perhaps the reasonable view is that mankind has to pay a price for the joy of the more exciting if more dangerous life provided by technology. It is up to society, presumably in the shape of the state, to institute whatever controls the people as a whole can agree to.

To come back for a moment to the accusation that technology is reducing the life of the individual to one of drab uniformity to an imposed standard in all aspects of life, let us look at the esthetic problem. The scientist and the technologist, being human, agree with the humanist that "man does not live by bread alone." Technology does not by itself create an esthetic, but it can be mightily influential in bringing one to vast masses of people. One would suppose that the sensible procedure for the humanist would be to take advantage of every technological improvement to propagate the great esthetic values which the race has built up over the ages. As a matter of fact, this is being done. We have already commented in Chapter 3 on the tremendous increase in appreciation of music which has been brought about by the manufacture and distribution with relative cheapness of high-fidelity recordings of classical music, accompanied by the production of relatively inexpensive sound reproducing equipment. All this would have been impossible without modern technology, which has thus brought the ability to listen to fine music at will within the range of millions. He would be an obscurantist indeed who would deny this technological contribution to the arts. At the same time, it must be admitted that it would be very desirable if there were greater collaboration between imaginative artists and technologists to take fuller advantage of the possibilities inherent in this contribution and to encourage more enlightened industries to pitch their radio and TV advertising programs on a higher esthetic level.

But does the enlightened humanist have a real point when he expresses his doubts about the beneficial results of modern technology on the life of people because it tends to make them more materialistic in their attitudes toward life and its problems? Are we being so overwhelmed with the gadgets which now play such a role in our lives that more fundamental matters are being crowded out? As some put it, is the machine taking over? Certainly more people than ever before are preoccupied with material things. We spend a lot of time thinking about our cars and the way they go. The same is true of our household appliances. And yet experience seems to indicate that when people gather, the conversation ultimately gets around to the age-old problem of the troubles of the individual. There is no sign that people are really losing the chief interest they have always had: other people. From this standpoint, the chief impact of the materialistic evolution brought about by technology is simply a widening of human experience. One can argue about this interminably. Some humanists plead for human interest to be restricted to the simple things, experience that nature provides without the intervention of man, the flowers, the birds, the bees, the stars, etc. But if history shows anything at all, it is that man has always striven to enlarge the scope of his experience; he is never contented with the simple life. Perhaps his strivings have brought him trouble and may bring him more, but who is to judge the rights and wrongs of this?

The moral seems to be that modern technology is a challenge to the intelligence of society to make the continual adjustment to the new experience being forced on mankind. It is here indeed that thoughtful people have developed their most serious misgivings. For man has always shown himself to be a warlike creature, and technology is providing him with weapons of devastating effectiveness beyond the wildest dreams of the early conquerors. This is without doubt a problem many orders of magnitude more serious than any other connected with the influence of technology on society. It would be nice if we could find a pat answer to it. It will not be solved in a hurry and not before serious harm has come to many people, though probably in the last analysis not to more in proportion to total population than in mankind's previous fruitless endeavors to solve human problems by the unpleasant plan of killing people. It probably will not be solved until the clever people learn to develop more meaningful scientific concepts about the behavior of man in society. One of the inevitable results of the adoption of a scientific attitude toward this problem will undoubtedly be the ultimate disappearance of the nation state.

Science and technology have had an interesting relation throughout recorded human history and in the physical domain, at any rate, are now so closely intertwined that there is no likelihood they will ever again draw apart. This strongly suggests that the same scientific method which is coming more and more to dominate the kind of technology which is being created should also be applied to see that the results of this technology are beneficial and not harmful to the human race. We might go so far as to say that modern technology is too subtle and complicated to be left to the scientifically ignorant to play with. This implies the optimistic belief that a genuine science of human behavior in society is not only possible but that it is in the making now. It will obviously not be a form of science isomorphic in all respects to that which has been so successful in the realms of physics and chemistry, but it will have to exhibit the same fundamental characteristics of accuracy of observations and description, zealousness and aptness in creation, and imaginativeness in understanding human experience.

CHAPTER 8

SCIENCE AND
THE STATE

Technological and Scientific Needs of Government

It is easy to forget that government has always had technological needs and perhaps natural to think that the complicated relations of science with the state came into existence only in very recent times. But even in the ancient city state, governmental activity could not be carried out without building structures of a public character. To be sure, these early ones tended to be religious buildings, but religion and government were from the first closely connected. The building of the pyramids and the tombs of the kings of Egypt was a vast technological undertaking which demanded supervision on behalf of the highest authority of the state. It is clear then that government has had an interest in technology from the very earliest historical times. The degree of sophistication naturally fluctuated with the elaborateness of governmental control itself. It was obviously more in evidence at the height of Roman civilization than in the disorganized feudal system which succeeded this in the middle ages, though even in the latter case evidences of it can be found.

Many other illustrations could be given of very early governmental interest in technology and science. The famous Hanging Gardens of Babylon, one of the seven wonders of the ancient world, were said to have been built by King Nebuchadnezzar (whose reign was from about 604-561 B.C.) to please his queen. It evidently took considerable technical planning and skill in design and construction and reflects, incidentally, the Babylonian competence in mathematics. One would give a good deal to know something of the background and status of the engineers who supervised this work. They might well have been military men, since one of the greatest technical needs of early government was weapons and methods of warfare. We shall explore this specifically in detail in the next section but for obvious reasons cannot fail to mention it here, since most governments of ancient times appeared to have maintained themselves largely through warfare, and heads of state were apt to be great soldiers.

This reminds us in particular of Alexander the Great, who was tutored in his youth by Aristotle and evidently imbibed thereby a love of learning and a realization of its importance for the ruler of a large domain. In the course of his conquests he collected or had collected for him by his officers large quantities of information about the topography, fauna and flora, and other characteristic features of the regions he conquered. It is believed that these helped to serve as a basis for much of Aristotle's scientific knowledge and what he taught in the Lyceum at Athens. On the deaths of Alexander and Aristotle, the famous educational center at Athens was closed, but Ptolemy Soter, the successor of Alexander in Egypt, made the city of Alexandria (founded by Alexander the Great in 332 B.C.) into a center of learning famous for centuries throughout the ancient world. He founded the famous library (about 298 B.C.) which flourished with vicissitudes until it was burned by the Saracens in A.D. 644. He and his successors also supported the Museum, a veritable ancient university with ultimately a hundred professors. It was here that Euclid (about 300 B.C.) taught and wrote his well-known elements of geometry, certainly a basis for a large part of both ancient and modern technology. Here also Aristarchus of Samos (third century B.C.) worked, the first apparently to propose the cosmological theory later associated with the name of Copernicus, i.e., the heliocentric theory. For many years the head of the library in Alexandria was Eratosthenes (third century B.C.), the great geographer and proposer of what was later adopted as the Julian calendar, a technological feat of great significance, for governmental activities cannot proceed with regularity without the existence of a reasonably accurate calendar. Nature somehow failed to make the rotation of the earth simply commensurate with its revolution about the sun, so the length of the year is not an integral number of days. Eratosthenes saw the necessity of interpolating a day every four years. Here of course he did not go quite far enough, but it was a marked technological advance. His map of the known world was of tremendous importance to the traders of his time.

From the standpoint of technology in the service of the state, probably the greatest student of the university of Alexandria was Archimedes. Most of his career was spent in his native city of Syracuse in Sicily, where it is usually stated he devoted himself to mathematical studies. It is an interesting fact, however, that he managed to become a very efficient consultant to the government of the city state, in the person of the king Hiero. The famous story is well known of his discovery of the method for measuring the relative amounts of gold and silver in the king's crown. This was of course connected with his great principle of the hydrostatics of bodies immersed in or floating on liquids, probably his greatest contribution to physical science. In the solution of this problem he constituted himself, so to speak, a one-man Syracusan bureau of standards. We have already commented on his discovery of the law of the lever. His justification is probably afflicted with fallacy. Nevertheless, there seems to be no doubt he had a deep understanding of the behavior of machines. His exploits in designing instruments of war have been celebrated by Plutarch, and in the war against the Romans he constituted himself a sort of National Defense Research Committee and contributed mightily to the protraction of the siege, though the Romans ultimately won and Archimedes died in the sack of the town.

It is well known that science in the sense of the Alexandrian school did not flourish under the domination of Rome. Nevertheless, the Roman state made great use of technology. It had to do so in order to build and maintain a far-flung empire. This demanded ships and roads for transport, as well as docks, lighthouses and bridges. The Romans built them all, and their technical competence in road building, in particular, is attested by the fact that many of their roads exist in part today, two millennia, more or less, since their original construction. It is important to note that they were built to serve the state. Surprise is often expressed that the Roman government did not particularly encourage the development of new technological devices, in view of its vast use of technology. There will always be some mystery surrounding this sluggishness. The Romans in general were not scientifically imaginative; they took most of the science they used from the Greeks and fell back on the use of slaves for the source of energy in their technical achievements. Since practically all of the hard work was done by slaves, there was little incentive for their owners, the aristocracy of Rome, to exercise the imagination to invent new technical devices. Many of the slaves were clever and inventive enough to do this, but here again there was probably no inducement, since, human nature being what it is, there is little incentive to invent what you cannot personally profit from, and slaves could not expect to profit from their inventions, even had they had the facilities for trying them out. What Rome lacked was a strong middle class of free people who could obtain a suitable education and could gain enough leisure to develop new gadgets and processes. As it was, the Roman state made excellent use of the technology taken from the Greeks and other peoples and kept it from dying out.

This is not a history of technology in government, so we must skip over much ground and content ourselves with illustrations which show clearly the dependence of the state on science and technology.

Consider Leonardo da Vinci, the versatile genius already mentioned in Chapter 3 in connection with his artistic merits. It is not so generally stressed that in 1502 he became chief engineer for Cesare Borgia, the great but unscrupulous military leader and cardinal of Florence and Rome. In this capacity he traveled extensively in Italy as a kind of military consultant. He became successively a technical adviser to the Papal Mint and an engineer in the French court. A substantial fraction of his career was thus spent essentially as a governmental technologist.

To pass on to a later century, take as the next example the Fleming Simon Stevin of Bruges, better known by the Latinized version of his name, Stevinus. He is famous in physical science for making clear the way in which an inclined plane works and for calculating its so-called mechanical advantage. Because of his competence in mechanics, Stevinus was appointed director of land and water construction by the States of the Netherlands and later became Quartermaster-General of the Dutch Army under William of Orange (called the Silent, 1533-1584), the leader of the Dutch revolt against Spanish rule and founder of the Dutch Republic. As the leading military engineer of his time, Stevinus became an authority on fortifications and took care of the technical side of the struggle against the Spaniards. His contributions were not confined to mechanics, since he also advocated strongly the introduction of decimals in arithmetic, with application to coinage, weights and lengths, etc. Here he was far in advance of his time; though his propaganda ultimately bore fruit on the continent of Europe, the use of the preposterously complicated English system is still widespread. The life of Stevinus is another prime example of the tendency of an enlightened government to seek out clever and competent men to serve its technological needs.

One of the most demanding technical problems confronting governments has been the control of water, e.g., the regulation of rivers and the draining of marshlands. It is interesting to note that governments have not hesitated to employ the services of scientists to meet such needs. Thus Galileo, whose scientific achievements have been commented on in various places in this book, was at one time superintendent of the waters of Tuscany.

Accurate navigation demands the ability to fix the position of a ship at sea. The latitude is relatively easily determined by the height of the sun over the horizon at noon (though we may recall from Chapter 5 William Gilbert's proposal to use the inclination of the magnetic needle for this purpose). Longitude, however, provides greater difficulty since it demands the ability to determine the local time at sea of some event and compare it with the standard time of the same event at an arbitrarily chosen meridian, e.g., Greenwich in England. Since governments on the seacoast have always maintained navies and have taken some responsibility for the safety of commercial shipping, the longitude problem was one that commanded a lot of attention and led to offers of prizes by various governments for its solution. It is interesting to note again that a scientist like Galileo was interested in this problem and devised a solution based on his discovery of the moons of Jupiter. Tables of the eclipses were drawn up, and it is clear that determination of the time of eclipse of a particular satellite at a position at sea when compared with the standard time for the same eclipse at the prime meridian will serve theoretically at least to determine the longitude. Galileo actually tried in 1616 to sell his method to the king of Spain for a considerable sum of money and a title. But his offer was not accepted. He had the same fate when he approached the States-General of the Netherlands towards the close of his life (1637). The method unfortunately was not very accurate because of the practical difficulties associated with the observation of the eclipse on a moving platform like a ship.

Of course, an obvious method of determining position is by "dead reckoning," based on knowledge of direction of motion of the ship and its speed. The so-called log was, however, for long a very unreliable measure of speed, and observations of the heavenly bodies seemed to offer the best hope of success. At any rate, this was the English view in the seventeenth century, for Charles II founded the Greenwich Observatory in 1675 chiefly for the purpose of obtaining better data on the motion of the moon, so that this could serve as a precise means of determining longitude. He actually wrote this into the instructions of Flamsteed, the first Astronomer Royal. Here the technological needs of government had a vital influence on the direction of astronomical research. Newton, of course, spent much of his scientific effort on the precise mathematical determination of the moon's motion. However, though doubtless better than Galileo's scheme using Jupiter's satellites, the moon method proved insufficiently accurate for the demands of navigation. The British government in 1712 offered a prize of up to £20,000 for the development of an accurate method for determining longitude. It was finally awarded after many delays to a Yorkshireman named John Harrison, who by 1764 had constructed a new type of clock called a chronometer which would keep time correct to fifteen seconds in five months. The ship's chronometer solved the longitude problem until the twentieth century, when radio time signals have provided nearly instantaneous time information at any place on the earth.

With the rise of sophisticated nation states in the seventeenth and eighteenth centuries and the growing realization that the welfare of the state depends on the economic well-being of the citizens (or at any rate a part of them), governments naturally began to pay more attention to technology. Thus we find France under Louis XIV and Louis XV endeavoring by governmental action to encourage more efficient methods of manufacturing as well as a more vigorous merchant marine. To give a push to the textile industry the famous chemist C. L. Berthollet was made director of the state dyeworks.

In 1785 he introduced chemical bleaching of cloth and thus effected a revolution in the industry, which soon passed to England and other countries. Another illustration of the practice of the French government during the eighteenth-century period of enlightenment to employ celebrated scientists on government projects is provided by A. L. Lavoisier, one of the greatest names in chemistry. Among other practical contributions, Lavoisier improved the quality of French gunpowder, which may have contributed to the success of the armies of the French Revolution. Perhaps in the long run an even greater contribution was his secretaryship of the commission that inaugurated the metric system in 1791. Ironically, his services did not prevent his execution on the guillotine by order of the revolutionary tribunal: he had the misfortune to be involved in the old-fashioned system of tax collection and became moderately wealthy. It does not always pay for scientific men to get too closely involved with government; there are also contemporary illustrations of varying degrees of poignancy.

Undoubtedly, both Berthollet and Lavoisier were stimulated to important basic research in science because of their technological labors for the state. Another related example is that of Benjamin Thompson, Count Rumford, the American-born scientist who became founder of the Royal Institution of Great Britain. At the end of the eighteenth century he was in the service of the Elector of Bavaria (who indeed gave him his title) as War Minister. In this capacity he supervised the boring of brass cannon. From his careful observations of the heat developed in this process, he convinced himself that the caloric theory of heat, so widely held in his day, was untenable. His work thus led him to be one of the founders of the so-called mechanical theory of heat, the point of view which looks upon heat as a manifestation of the random motion of the constituent parts, the atoms and molecules, of matter.

It is a little difficult and perhaps for our purposes not wholly necessary to separate the efforts of the state to satisfy what it considers to be its own vital needs through science and technology from its desire to promote these activities for the benefit of both individual citizens and the mass. We would be remiss, at any rate, if we were to overlook a significant governmental method for promoting zeal in technological invention. This is of course the idea of a patent, which in brief confers on an inventor the right to exploit his invention to his profit in return for his willingness to disclose the details to the state and the public. Human nature being what it is, the average clever person who hits upon a novel idea and has enough technical competence to turn it into a useful gadget desires to make money out of it before other persons exploit the scheme. His tendency is therefore to keep the whole thing as secret as he can so that his process may not be copied. But it is clear, as was early recognized, that this behavior does not encourage the progress of technology in general, since it is often the close study of the results of the endeavors of others which stimulates an inventor or scientist to a new idea of his own. So the notion of a patent right arose, presumably first in Venice in the latter part of the fifteenth century in connection with glass-making. The idea spread from Italy to England and other European countries. In England the first patent dates from 1552. Obvious abuses followed, such as the conferring of monopoly rights in return for money payments to the Crown. There have been numerous patent laws extending down to the twentieth century. According to the law now in force in the United Kingdom, the normal length of a patent is sixteen years, though this may be extended if the inventor can demonstrate hardship. The importance of patents in the eyes of the founders of the United States was so great that they incorporated among the powers of the Congress in Article I, Sec. 8, of the Constitution the following provision:

> The Congress shall have power to promote the progress of science and useful arts, by securing to authors and inventors the exclusive right to their respective writings and discoveries.

This provision has been made the basis for numerous copyright and patent laws. The United States Patent Office was created in 1790 in order to pass on applications for patents or inventions. By 1860, 36,000 patents were granted, but this gave very little indication of the enormous activity to come, for in the period from 1900-1925, the total of patents granted reached 969,428! American law allows a patent to run for seventeen years. In very recent years there has been a decline in the number of patent applications. It is perhaps a bit premature to seek for a definitive reason for this. Some find it in the fact that most inventions now are not made by independent inventors but rather by employees of large industries. Most patents involving ideas basic for large-scale manufacturing are subject to extensive litigation which only a big concern can afford. This may well change ultimately the picture of government assistance set forth in the quoted provision in the Constitution. The subject is a very complicated one, but in this respect is typical of the direction in which practically every relation between government on the one hand and science and technology on the other has evolved.

Only the very simplest technology can get along without measurement. People were measuring land, for example, several thousand years ago, long before there was much in the way of science. But measurement demands units, and these in turn must be subject to some degree of standardization, or the whole activity becomes meaningless. The use of parts of the body (length of forearm, for example) for units of length was at first natural, but it was early realized that people differ and someone must of necessity set the standard. The only agency powerful enough to adopt, maintain and enforce such a standard was the state, and this function of government has persisted from the earliest times to the present. As technology became more sophisticated and more affected by the advancement of science, the need for standardization in measurement became ever more pressing, as the tendency spread for individuals and private industries to adopt their own standards with respect to new products and processes. Modern governments have responded to this situation by establishing laboratories whose principal function it is to maintain standards of measurement. In Great Britain (conservative, as usual!) the need was first met, indeed, by a committee of the British Association for the Advancement of Science. This organization was founded in 1831 by a group of scientists headed by Sir David Brewster, the Scottish physicist and biographer of Newton. Among its main objects was "to promote the intercourse of those who cultivate science with each other." From the first, it took a lively interest in the relations of science with technology. There was a strong realization that if science and in particular the new electromagnetism was to be applied successfully, satisfactory units would have to be established for electrical quantities and fundamental constants carefully evaluated. The Association appointed a Committee on Standards to take care of these activities. These became so multifarious that as a result Great Britain established in 1902 the National Physical Laboratory (Teddington, Middlesex) to devise, maintain and improve standards for quantities of technological importance. In the United States the same purpose was fulfilled by the establishment in 1901 of the National Bureau of Standards in Washington. As a matter of fact, both Britain and the United States were behind the continental European nations in establishing standards laboratories. Thus, in 1875 there was set up in Sèvres, a suburb of Paris in France, the International Bureau of Weights and Measures to maintain and extend metric standards for all countries using the metric system. This arranges for frequent international conferences on standards. The United States was a signatory to the international treaty which established this Bureau and sends representatives to the conferences. It is well to recall that though English units generally prevail in America, they are all defined in terms of the corresponding metric standards.

Germany was not far behind France in establishing a federal bureau of standards in the form of the Physikalisch-Technischen Reichs Anstalt. This was established in 1887 in Charlottenburg near Berlin.

A few words about the National Bureau of Standards of the United States will serve to describe the historical development and contributions of institutions of this type. Though the Constitution of the United States had given the Congress the mandate to fix standards of weights and measures and though there had long been in existence in the Treasury Department an Office of Weights and Measures, by 1900 the latter had become wholly inadequate to meet the demands of rapidly developing industry. The abolition of this office and the establishment of the Bureau of Standards were necessary steps in the assumption of genuine governmental responsibility for setting standards in the light of contemporary requirements. Progress of the Bureau's program, which was inaugurated under the direction of Dr. Samuel W. Stratton, a physicist who had previously taught at the University of Chicago, was rather slow, since it was necessary to construct an entirely new set of laboratory buildings and at the same time face the problem of establishing new standards for the host of physical quantities emerging from the progress of science.

The Federal Act of March 3, 1901, which set up the Bureau, was amended extensively in 1950 and gives the institution considerable latitude in the performance of its duties. The mission of the Bureau as recently (1958) stated by the Director is as follows:

A. To provide, as a Federal responsibility, the central basis for a complete consistent system of physical measurement of national scope adequate for the expanding national activity in scientific research, engineering and commerce, and properly coordinated with similar systems of other nations.

B. To do research and engineering in other areas for which NBS has an assigned central Federal responsibility. Examples are: radio propagation, data processing systems, cryogenic engineering, building technology, and fire research.

This is clearly a very broad mission. It has received its interpretation in the five classes of activity carried out currently by the Bureau, briefly summarized as follows: (1) creation and maintenance of standards of important scientific quantities; (2) methods of measurement and dissemination of standards; (3) measurement of properties of matter; (4) supporting technical services; (5) cooperative and advisory services. These activities are carried out in twelve scientific and technical divisions covering most of the principal branches of physics, as well as chemistry, mathematics and engineering. Each division is further divided into sections having duties confined to a specialized field. Thus, the atomic and radiation physics division has sections devoted to spectroscopy, radiometry, mass spectrometry, solid state physics, electron physics, etc. Two observations must be made. Though the Bureau functions as a standardizing agency, in order to develop standards it must also undertake basic research, and it does this on the same scale and thoroughness as a university scientific laboratory. To this extent it functions as a great national scientific laboratory, with emphasis of course on the physical sciences. The second observation is that since science and its branches are continually evolving, and in particular since the relations between the various branches are changing rapidly in the contemporary scene, the organization of the NBS must also change frequently. This is misunderstood by some people, who fail to recognize the rapidly developing character of modern science. The National Bureau of

Standards has attained a position of great eminence among the scientific laboratories of the nation.

We shall speak of other governmental laboratories and research institutions in subsequent sections.

Science and National Defense

Warfare is apparently as old as the human race, and groups and nations have always felt themselves justified in using any means that came to hand in defending themselves against those they considered their enemies. The link with technology is obvious, and as we have already stressed in the previous section, a large segment of the technology developed by the human race has been concerned with weapons of war. It is not necessarily true that every weapon was invented primarily for use in warfare; it is likely, for example, that the bow had its first use in hunting animals for food. But from this, the transition to its employment against human beings was all too easy. Similarly, gunpowder was presumably hit upon by accident by the Chinese before A.D. 1000 in their experiments with fireworks: certainly it was used by them for a long time for this purpose. It is usually supposed to have been invented independently in Western Europe in connection with alchemical investigations. In any case, clever people soon sensed its possibilities in warfare. By A.D. 1300 it was being used in cannon to replace the spring-powered catapults and other similar devices for hurling large objects.

One of the most dramatic episodes of the early use of technology in war concerns the defense of Byzantium or Constantinople against the Saracens. In A.D. 712, the latter made one of their most determined assaults on the city, coming up the Bosporus with a fleet of some 1800 ships and galleys, according to the early chroniclers. The odds against the defenders seemed very great indeed, but a new form of naval tactics made its appearance. Instead of dispatching galleys to grapple with the foe in the approved fashion of ancient sea warfare, the Byzantine Greeks now relied on fire ships which hurled fire balls into and around the enemy vessels. The result was disorganization and ultimate destruction of the Arab fleet. The fire used was no ordinary fire. A chemist named Callinicus originally from Egypt or Syria (it is uncertain which) had made the remarkable discovery that naphtha mixed with sulphur and pitch forms a highly combustible material, difficult to extinguish since it burns in water as well as in air. This was the celebrated Greek fire. No one can actually be sure who Callinicus was, whether he really invented the fire, or when; but there seems to be no doubt that its use in the defense of Constantinople proved decisive.

The invention and use of Greek fire in warfare provide another interesting facet of the connection between technology and military defense. Having found a useful weapon, the Byzantine Greeks did their best to keep it to themselves. Gibbon in his *Decline and Fall of the Roman Empire* is authority for the view that they actually managed successfully to guard the secret of Greek fire for some four hundred years, and in the meantime developed many new and ingenious methods for its more effective use, including copper tubes for the forward projection of the fire, as in the modern flame thrower. Eventually, the Arabs penetrated the mystery and adopted the use of Greek fire in their own military tactics. The moral of this is plain and has been substantiated again and again in the course of military history: no technological military secret has ever been successfully kept from the prying eyes of the enemy or likely enemy, and the time lag before rediscovery has steadily shrunk. The "cloak and dagger" mystery stories usually stress the view that the secrets are "stolen" by espionage, but as military technology has come to depend more and more on basic science, it is clear that the dissemination of such secrets has become largely a matter of actual rediscovery. Science knows no national boundaries, and no nation has ever had a monopoly on the ability to produce successful scientific ideas.

So far we have been talking mainly of technology in warfare, though the preoccupation of scientists like Archimedes, Leonardo da Vinci, Stevinus, Lavoisier, etc., with military matters at certain stages of their careers suggests that basic scientific ideas can enter vitally into the technological development of a weapon or a military tactic. It remains true, however, that the very close connection between science and technology which we have explored in the previous chapter and which began to become obvious only in the early part of the twentieth century did not manifest itself in warfare on a large scale until the Second World War, from 1939-1945. Compared with the magnitude of the role of science in war during this period, the efforts in World War I (1914-1918) were merely tentative, though some interesting innovations were suggested and tested. Thus, it was pointed out by scientists that the position of heavy guns could be fixed by acoustical means. The sound of the explosion of the gun reaches each sound-receiver (or microphone) of an array of such receivers at a different time. From an analysis of these time differences, the location of the gun can be determined. The method received practical and reasonably successful use during the later stages of World War I and was even more extensively employed in World War II.

The devastating effect of submarine activity during the First World War led to the attempt to devise successful schemes for their detection. It was early recognized that electromagnetic radiation, whether in the form of visible light or as the longer radio waves, is absorbed so rapidly in sea water that its use as a detection means is minimal. On the other hand, it had been known for a long time that sound travels through water in general even better than through air. It was hence suggested that sonic listening devices called hydrophones could serve to locate submarines and other submerged noise producers in much the same fashion in which guns were located by microphones. These were tried and found to be moderately successful. The first hydrophones were merely tubes with hollow rubber spheres attached at the ends submerged in the water. They were not very sensitive and were ultimately replaced in the first instance by the carbon granule type of microphone similar to that used in a telephone transmitter and later by the so-called piezoelectric receiver. The latter was developed by the French physicist P. Langevin, based on the discovery by the brothers Pierre Curie and Jacques Curie in 1880 that certain crystals (e.g., quartz) have the property of developing electric charges on certain faces when deformed by stresses and, conversely, when placed in an oscillating electric field, will vibrate with a definite frequency related to that of the electric oscillation. Such crystals have been called piezoelectric. For a long time their properties were only a scientific curiosity, but the invention of methods of producing and maintaining electrical oscillations of high frequency and high power soon made it possible to employ the piezoelectric effect in the construction of very intense ultrasonic sound sources (cf. page 138). The same crystals also are available as highly sensitive sound receivers.

In the period between the world wars all the great naval powers developed underwater sonic and ultrasonic devices based on the scientific principles just mentioned as well as others (e.g., the so-called magnetostrictive effect, in which the act of magnetizing a bar of ferromagnetic material produces a slight change in its length) into which it is not necessary to go. In order to avoid dependence on the sound output of the object to be detected, an echo method was devised, known as sonar. In this, a concentrated beam-like pulse of high-frequency sound is projected into the water by a submerged source and either by mechanical or equivalent electrical rotation scans the environment. When the pulse encounters a solid obstacle, an echo is returned and is picked up by a sensitive receiver mounted near the source (sometimes the source itself is used as a receiver, on an on-off basis). The bearing of the obstacle (e.g., submarine) can then be determined, as well as its distance (from the time between the emission of the pulse and the return of the echo). Sonar has proved to be a very significant instrument in undersea warfare, and provides an excellent illustration of the use of science via technology in national defense.

A few observations are in order. It would be easy and natural to suppose that once the utility of the scientific principle was clear, the rest of the story would be entirely a matter of technological ingenuity, i.e., a job for the engineers to design and construct the appropriate equipment with maximum efficiency, stability and ruggedness under operating conditions. This is of course true to a certain extent, but the fact that it falls short of the whole truth has decisive significance for the whole problem of the role of science in warfare. Actually, it was found that even with thoroughly tested and calibrated equipment sonar results in ocean water were often thoroughly unreliable. The suggestion was strong that a more careful scientific study of the medium of propagation should be undertaken. This led to a much better understanding of the nature of sea water and hence contributed to the science of oceanography. It also led to discovery of deep sound channels available for the long-distance transmission of underwater sound.

Many other illustrations of this same general situation could be cited. They have led defense agencies of the United States government, as well as foreign governments, to put a good deal of emphasis on the performance of fundamental research as a basis for new weaponry. As military personnel often like to put it: "You cannot expect to fight the next war with the weapons of the last one." And new weapons inevitably demand the sort of imaginative approach which only basic science can provide. Some have questioned indeed the effect that preoccupation with defense research may have on fundamental scientific research. Undoubtedly, some areas suffer neglect, especially during the intensive activity in actual wartime. But even here it is possible to distort the picture if one is not apprised of all the facts. Modern weaponry, thanks to the imagination of scientists, has become so complicated and sophisticated that there is scarcely any branch of science which is not involved in it to a greater or lesser degree. And, as has already been suggested, one rarely gets very far with any weapon without discovering the need for further scientific research.

In the period of the Second World War and since, the most striking example of science in warfare is of course the development of the so-called atomic bomb. Even the possibility of this kind of explosive was not dreamed of before the discovery of radioactivity by Becquerel in 1896, and though the gradually increasing knowledge of the twentieth century suggested that the binding energy of the nuclei of atoms (a twentieth-century scientific concept in itself) was relatively enormous, even as late at 1920 very competent atomic physicists expressed the strong conviction that no artificial way of releasing this energy would ever be practical. Yet, further fundamental research in the new field of nuclear physics led to the discovery of neutron-induced nuclear fission in 1939 and the practical application to the construction of the bomb followed rapidly. One reason for the relatively short time that elapsed between the scientific discovery and the technological success is of considerable importance for modern scientific warfare and should therefore be given some emphasis. This is the role played by the cooperative activity of large groups of scientists and engineers. Many thousands worked on various phases of the atomic bomb, each group concentrating on a particular problem. Under skillful direction and careful coordination, this plan was so successful that a feeling grew up in certain quarters during and after the war that teamwork was the key to success in scientific research and technological development. No one can doubt the tremendous advantage of cooperative effort in such development. Quite another thing, however, is involved in scientific discovery, and it would probably be unfortunate if the feeling became prevalent that a valuable new recipe for obtaining basic scientific results had emerged from this technological success. The vital role of the individual thinker in scientific advances can scarcely be overestimated, and any attempt to make the original researcher fit the pattern of a team with its hierarchy of responsibility may well lead to disaster.

The extent to which contemporary science has led directly to new weaponry in the past quarter century has been so vast that an attempt to provide even an outline would be unrealistic.[1] Everyone has heard of the research on the atomic bomb as well as the radar which was so decisive in detecting enemy aircraft. Well known also is the proximity fuse, a device which controls the firing of a warhead in terms of its nearness to a target. The homing torpedo, in which the sonar idea is used to guide the torpedo directly to its target, is another highly technical weapon.

Less well known but of tremendous importance in modern warfare is the purely "paper" scientific application called "operations research." This has been defined[2] as "a scientific method of providing executive departments with a quantitative basis for decisions regarding the operations under their control." A military illustration will serve to clarify this. During World War II, in order to minimize submarine action, Allied merchant shipping proceeded in convoys. The amount of escort protection was still limited (approximately six escorts to forty ships) and losses ran high. It was decided to study statistically the relation connecting the various factors entering into the problem: the number of losses per convoy per

[1] The reader interested in details will find it worthwhile to consult the following books, more or less popular in character: J. G. Crowther and R. Whiddington, *Science at War* (New York: Philosophical Library, 1948); H. D. Smyth, *Atomic Energy for Military Purposes* (Princeton University Press, 1945); E. W. Titterton, *Facing the Atomic Future* (London: Macmillan, 1956); J. F. Baxter, *Scientists Against Time* (Boston: Little, Brown & Co., 1946).

[2] P. M. Morse and G. E. Kimball *Methods of Operations Research* (New York: Technology Press of M.I.T. and John Wiley and Sons, Inc., 1951).

trip, the size of the convoy, the number of protecting escorts, etc. A very interesting result emerged from this study, namely, that the absolute number of ships lost per convoy per trip was practically independent of the size of the convoy. This at once suggested that to cut down the *relative* or percentage loss, the number of ships in the convoy should be made as large as possible—a rather simple conclusion, but one that, when adopted, made shipping much more efficient in the prosecution of the war effort.

One might object that if this is all there is to operations research, its importance is much exaggerated and it is simply common sense. As a matter of fact, all decision-making based on a careful study of all relevant factors in a given situation is really operations research and as such has been going on since time immemorial. However, it is only in recent times that it has been analyzed formally and put on a mathematical basis. In most problems to which this analysis has been applied, statistics and probability considerations have had to be freely used. No defense agency any longer develops a weapon or a counterweapon without studying its uses from the operations research point of view. Of course, operations research is not restricted to military activities: it is applicable to any complex business operation, such as traffic control and quality control in manufacturing.

Science in defense does introduce some problems not evident in the pursuit of scientific research in a university. The most important of these involves the factor of security. We have already commented on the natural tendency for the possessor of a successful weapon to try to keep its design and method of construction and operation secret from a possible enemy. When science enters importantly into weaponry, it is altogether too easy to extend this attempt at secrecy to the basic scientific principles whose applications have made the weaponry possible. This attitude and emphasis is of course the very antithesis of the spirit of science, which thrives on open communication among scientists everywhere and withers in isolation. Moreover, as we have seen, the attempt to keep scientific ideas secret is foredoomed to failure. It not only does not work, but it exercises a very unfortunate retarding influence on the scientific progress of those who practice it.

Overzealous classification of research reports as confidential or secret has often hampered the carrying out of effective research on military projects, since it has served to keep certain information from those who needed it in the prosecution of their work. This has in turn led to the needless repetition of costly experiments to secure data actually locked up in secret reports. World War II provided many such cases, even to the faintly ridiculous situation in which a man who had prepared a report, which was classified secret, was later refused the right to examine his own report. Declassification was supposed to solve such problems, but usually this process proceeds so slowly as to be useless. A possible solution would be to put the declassification in the hands of scientists and take the job away from the military. However, this would only waste the time of valuable people. The only sensible answer would appear to be to refrain from security classification of any purely scientific report and restrict such to reports on actual details of weapons. It is unfortunately unlikely that this ideal solution will be attained in the foreseeable future. The problem will continue to remain one of the most serious obstacles to the successful use of science in national defense.

Where should scientific research for defense be carried out? In the United States, the Department of Defense maintains many laboratories for research and development. Thus, the Department of the Navy has the Naval Ordnance Laboratory (Silver Spring, Maryland), the Naval Research Laboratory (Washington, D.C.), the Navy Electronics Laboratory (San Diego), the U.S. Navy Underwater Sound Laboratory (New London), etc. The Air Force maintains extensive laboratory facilities at Wright Air Force Base in Dayton, Ohio, and at the Air Force Cambridge Center in Bedford, Massachusetts. The Army has the Aberdeen Proving Ground (Aberdeen, Maryland) and the Signal Corps Laboratories in Fort Monmouth, New Jersey. All told, these employ several thousand professional scientific personnel and many more thousand technical assistants. While most of the work that goes on in them should properly be termed development, e.g., the design of weapons and weapon systems, they also perform much basic research, for example, on the acoustic properties of sea water and the influence of atmospheric properties on the propagation of radio waves. In general, of course, the basic research is related rather closely to the needs of weapons and weapon use. In certain laboratories an enlightened directorate makes possible the pursuit of fundamental research in the fashion of the universities.

In addition to supporting research in their own laboratories, the various defense agencies contract research to private laboratories, both university and industrial. Thus, for example, the Office of Naval Research finances basic research in acoustics in some half dozen large university physics departments and in addition has much larger programmatic contracts in the same and other institutions. These latter projects involve special attention to Navy needs, e.g., the effect of bottom reflection and absorption in shallow-water sound transmission. On the whole, it is questionable whether such programmatic research is satisfactory for the universities, particularly if it is subject to security classification. Much government-sponsored research in the universities is worked on by graduate students under the supervision of faculty members, and the tendency has grown up to permit students to count some if not all of such research toward the doctoral thesis. This becomes complicated if the research is classified, since doctoral dissertations should be freely publishable.

Few now question the desirability of the support of scientific research in universities by defense agencies, but many difficult problems would be avoided if this support could be confined entirely to basic research in areas generally admitted to be promising for future weapon developments. On this plan, all classified research and development would be carried out in the defense laboratories. This would not preclude close liaison between these laboratories and the universities. In fact, on any scheme for the employment of science in national defense, such a connection is necessary, and it must be close. Nevertheless, it is clear that the educational responsibilities of the universities cannot be met satisfactorily if too much faculty time is devoted to programmatic research. While science faculty members must do research in order to keep themselves abreast of the times and stay academically alive, they must also teach if the education of the young is not to go down the drain, and it is becoming increasingly clear that in the United States considerable interference with teaching duties on the part of senior professors is developing because of preoccupation with the administration of governmental research projects. The problem is a difficult one due to the shortage of trained scientific manpower, but the suggestion made above would be a step toward its solution.

Financial Support of Science by Government

In previous chapters and notably in Chapter 7 it has been made abundantly clear that contemporary basic scientific research is an expensive business. Apparatus for experimentation in the physical sciences tends to be elaborate and costly, as most people realize when they read in the newspapers figures on the cost of "atom-smashers" and similar devices, which run up in the many millions of dollars. It is true that much very useful and even exciting physical research is being done with relatively modest sums, but the modesty is measured only by the fact that the cost might be of the order of thousands rather than millions. The life sciences are rapidly increasing their use of complicated physical equipment and hence contributing toward the increased expense. The sums involved are at any rate beyond the resources of most colleges and universities. Even theoretical research is expensive, since it demands adequate library facilities, and books and journals continue to rise in price. The question arises: who is paying for all this basic research and who will assume the growing obligation for the future? There seems to be but one answer: government. We have indeed commented in Chapter 7 on the role of industry in the support of scientific research, but it is clear that most of this is more properly labeled applied research and development.

Let us look at the extent to which the government of the United States finances scientific research. We can get perhaps the best view of this from a publication of the National Science Foundation,[3] one of the duties of which is to conduct a continuing program of surveys of research and development in this country. For simplicity we shall first restrict attention to the support of research and development in the colleges and universities. In the fiscal year 1953-1954, the total sum expended in this sector (including research supported in so-called research centers associated with but not an integral part of the educational institution, as for example, the Lin-

---

[3] *Reviews of Data on Research and Development,* Report No. 16, NSF 59-65 (December, 1959). See also Report No. 36, NSF 62-32 (September, 1962).

coln Laboratory of M.I.T., the Stanford Research Institute at Stanford University, etc.) came to 450 millions of dollars. Of this sum the educational institutions themselves contributed 130 millions or 29 per cent, while the various agencies of the Federal Government contributed 280 millions or 62 per cent. Industry and other nonprofit institutions (private foundations and the like) made up the balance. Evidence shows that in the years since 1954, the percentage of support from government is growing slowly but steadily relative to the other sources.

Figures on research and development in industry tell the same story. In 1953-1954 direct industrial support for these activities amounted to 2200 millions of dollars, whereas the agencies of the Federal Government contributed 1430 millions, the percentages being approximately 60 per cent and 40 per cent, respectively. The corresponding figures (unrevised) for 1956-1957 were 3260 millions from industry and 3228 millions from government, or percentagewise about 50 per cent each. The trend is unmistakable.

It must of course be remembered that most of the financial support just alluded to went for development or applied research and a relatively small amount for basic research. Moreover, most of the development has been devoted to defense requirements. We have already commented in Chapter 7 on the relatively small fraction of total governmental and industrial research funds which find their way to the support of basic research. Nevertheless, in 1950 recognition of the importance of science in the national life led the Congress of the United States to set up the National Science Foundation "to promote the progress of science." This was a development of great significance, for it was clear from the outset that it would be the primary task of the new agency to do all in its power to stimulate basic science in the United States. Broadly speaking, the Foundation has sought to discharge its mission by (1) disbursing funds to support fundamental scientific research in colleges, universities and other research laboratories; (2) disbursing funds to support graduate fellowships, both predoctoral and postdoctoral, in the universities; (3) carrying on a continuous survey of the progress of scientific research in the United States; and (4) disbursing funds to improve the teaching of science in the secondary schools and institutions of higher education in the United States.

In the first activity just mentioned, the Foundation entertains research proposals from any scientist connected with a recognized educational or research institution. This may be a university or it may be a research organization like the Woods Hole Oceanographic Institution or the Mellon Institute of Industrial Research. By far, the larger number of proposals come, of course, from universities. Each proposal must outline a more or less definite program of research in a specific field and provide evidence of the competence of the proposed investigators as well as the interest of the institution in the work and its ability to provide fundamental facilities, e.g., space and utilities. A somewhat detailed budget must accompany the proposal, analyzing the proposed expenditure into such categories as salaries (or portions thereof) of the principal investigator and assistants (e.g., graduate students), and cost of material, equipment and supplies. These proposals are examined by advisory panels in the various disciplines who take advantage of advice received by expert referees throughout the country. Recommendations for awards are then made to the Director and the National Science Board. In the early years of the Foundation's activity, when the total budget was relatively low, approximately one in four of the scientifically approved proposals led to actual grants. More recently, with the increase in the budget, this fraction has been raised to about one in two. The Foundation naturally aims ultimately to make an award to all whose proposals are scientifically approved. Though most of the awards go to faculty members in universities where research is already well established, there has grown up more recently a tendency for the Foundation to make research grants to professors at colleges of liberal arts and sciences, where research tends to be weaker and where graduate study, if carried out at all, is limited to work for the master's degree. This is part of a program to strengthen science in the institutions which feed students into the larger graduate schools. It is based on the assumption that the colleges find it difficult to recruit good science teachers unless the latter are offered opportunity to carry out research, the support of which is usually beyond the financial capabilities of the institution's own budgets.

Though in the charter of the National Science Foundation the term science is interpreted broadly enough to cover the social sci-

ences as well as the natural sciences, in actual fact little aid was extended to the former in the early days of the Foundation. Since about
1957 there has been a developing interest in the support of research
in the history, philosophy and sociology of science as well as in such
fields as mathematical economics and anthropology. Presumably and
hopefully this will increase in the future, though it is pretty well
known that the Congress does not look upon this with the same enthusiasm with which it views research in the natural sciences, as
having more direct presumptive connection with the industrial and
military progress of the United States. One can glimpse here without
uncommonly acute perception one of the difficulties associated with
the support of science by government: the people responsible for
the appropriation of the taxpayer's money are apt to take a somewhat limited view of the nature of science. The need for great persuasive skill on the part of the directorate of a government agency
like the National Science Foundation is clearly evident.

The second important activity of the Foundation is the disbursing
of funds to provide graduate fellowships in the universities. Here
again, the aim is to promote the progress of science by insuring the
education of potential scientists. The Foundation supports several
different programs, some for predoctoral candidates and others for
postdoctoral. These fellowships are awarded on the basis of ability
only, as judged by panels of university scientists who study the submitted credentials. The stipends awarded compare favorably with
normal university graduate school fellowships and tuition is also
paid by the Foundation. A cost of education allowance is made to
the institution where the student resides, in recognition of the fact
that tuition never completely covers the cost of education. In addition to the postdoctoral research fellowships, the Foundation also
awards so-called Science Faculty fellowships to permit college teachers (mainly in the smaller colleges) to take leave of absence and
spend a year at a larger institution to refresh themselves and hence
hopefully become better science teachers. All told, the Foundation
provides several thousand fellowships each year. Though it is a bit
early to give a complete appraisal of the results, there is no question
that the education of scientists in the graduate schools of the United
States has been tremendously stimulated by the program. Some
pessimists, indeed, have questioned whether there is really enough

good scientific potential in the country to justify the effort and financial outlay. On the other side, many humanists have complained bitterly over the preferential treatment of science education which the program entails. This has gone so far as to lead to a serious proposal that the government set up a National Humanities Foundation to provide similar help in the education of nonscientists.

In connection with the third task of the Foundation we have already had occasion to refer to the Foundation's *Reviews of Data on Research and Development,* of which thirty-six had appeared through September, 1962. These summarize numbers and distribution of scientists throughout the country in various fields, as well as funds available from all sources, governmental and nongovernmental, for the support of scientific research. To keep track of all this is a monumental job, but there can be no meaningful study of the sociology of science without the data so accumulated.

The final form of activity of the Foundation relates to the improvement of the teaching of science in the secondary schools so as to insure the better preparation of students who go on to college and to give those who do not a better understanding of the nature of science, which will continue to have such a profound influence on their lives. The program has entailed setting up summer institutes in science at colleges and universities attended by high school teachers who refresh themselves in their respective subjects. To this plan there has been added the academic year institute, in which teachers on leave from their schools are supported for a whole college year and may carry on enough work to secure a master's degree. Millions of dollars per year have been spent on this program and much has been hoped from it. Yet it remains one of the more doubtful of the worthwhile projects for the government encouragement of science: the essential reason is to be found in the rather unhappy state to which high school teaching in science and mathematics has fallen in the past twenty-five years. So many of these teachers have had such poor collegiate preparation in the materials to be taught that they have found it very difficult to take full advantage of graduate courses, especially in mathematics and the physical sciences. Most of them really need undergraduate courses. But school systems demand graduate degrees for advancement. So here the National Science Foundation has collided with a fundamental weakness in American education. No one really knows what the outcome will be. The hope is that some good will ultimately emerge from the program, considering the large financial investment involved and the rather desperate educational situation being attacked.

It is of interest to note briefly the magnitude of the operations of the National Science Foundation in terms of money expended. The total available for the operations of the Foundation during the fiscal year 1960 was approximately $159 million; of this, about $57 million was expended in support of basic research, the activity listed under (1), above. Some $13 million went for the support of fellowships. Educational institutes mentioned under (4), above, took $33 million. Dissemination of information about scientific research claimed about $5 million. The rest of the total outlay was accounted for by the expense of running the Foundation and taking care of numerous special projects.

The National Science Foundation is not the only Government Agency (aside from defense agencies already discussed) which disburses large amounts of money in support of science. The National Institutes of Health also have huge sums at their disposal for the promotion of research and education in the life sciences. Public health is of such general concern that it is not difficult to arouse governmental interest in it. We need not go into details, since the story is similar to that already told about the National Science Foundation, though with its traineeship grants to universities the NIH is endeavoring to combine education and research in a way somewhat different from that of the NSF.

All this governmental outlay since the end of World War II on research and education in science has had an enormous impact on the tempo of research activity and has led to the vast expansion of scientific departments in most universities. It is gradually, though more slowly than might have been expected, pushing up the output of science Ph.D's. from these institutions. All this has not been without its problems. Curiously enough, the major one, about which many very sincere persons worried most when increased support to scientific research in the universities was first being discussed, has not materialized to any extent. We refer here to the natural fear that government financial support inevitably leads to governmental control. So far as can be discerned, no educational institution is being subjected to pressure with respect to any of its educational policies because of the receipt of governmental funds in support of some of its programs. To be sure, the expenditure of grant funds is subject to audit just as are the institution's own funds. This is usually considered a nuisance, but it can hardly be thought of as constituting government control. More of a problem has been the negotiating over a reasonable overhead to cover indirect costs. It is not surprising that government fiscal officers rarely see eye to eye (at first glance, at any rate) with university business managers and controllers on this point. University administrators quite naturally feel that in accepting a government contract to do research they are providing physical facilities which would cost the government very large sums to produce from scratch and that therefore a substantial financial allowance should be made for the existence of these facilities and their effective maintenance in addition to the direct cost of the research program, i.e., salaries for personnel and money for equipment, etc. The general principle of this has been readily admitted by government for contract research, at any rate, but the detailed figures have proved pretty sticky to agree on. When research has been supported by direct grants (as has been the plan adopted generally by the National Science Foundation) the attitude has been to consider that this research would be done anyway by the institution if it had the money and hence an indirect allowance should be kept to a minimum. Universities have protested, and as a result some increases in overhead have been allowed. It is possible that government agencies are a bit suspicious about the accuracy of the general accounting in private educational institutions, and some think this suspicion is not wholly unjustified. As a rule, universities are not run like business organizations and probably never will be.

The main point of our considerations here is that the problems involved in government financial support of scientific research in the universities have not been the much feared ones of government control but rather the day-to-day bickering involved in financial relations between any two parties anywhere, even with the highest ideals and best intentions.

More serious than the arguments over precise financial arrangements has been the short-sighted attitude of government toward the need for expanded major facilities, i.e., actual buildings to house new and expanded research programs. Somehow it was overlooked that no one in his senses can expect continually to crowd new research projects involving vast increases in personnel and elaborate equipment into old laboratory structures without overcrowding them to the point of hopeless inefficiency. But this is what has happened. Somehow it was expected that Santa Claus would provide the buildings while the government provided the money to run the research projects in them. The exception indeed was the creation of the great national nuclear laboratories like Los Alamos, Brookhaven, Argonne, and Oak Ridge from funds of the Atomic Energy Commission. But the rule was definitely against allowance for building, and the results have been unhappy. However, perhaps the story will have a happy ending, for it now appears that governmental agencies are to be permitted to disburse money for the enlargement of present laboratory facilities and the erection of new ones. Relatively small amounts have so far been allocated for this purpose, but a start has been made. Naturally, the problem is a complicated one, since the question of title to such structures arises. This immediately raises the whole thorny issue of direct government aid to higher education. It will doubtless be some time before this is resolved. In any case, it transcends the special matter discussed in this section.

To sum up, government support of science in the United States at every level of sophistication and complexity is here to stay and will grow indefinitely.

Freedom of Science

In the long pull perhaps the most fundamental problem posed by the increasing involvement of science with the state is that of freedom. Can a science which owes so much of its financial support to governmental subsidies remain really free? But what shall we mean by freedom in science? In an absolute sense we shall take it to signify the freedom of every scientist to pursue any line of investigation he chooses within his means or means which he can persuade others to supply, and to report his results freely in scientific journals for all the world to inspect. It is here assumed that the fundamental urge of the scientist is his curiosity about the world of his experience. But even if he is motivated by the desire to gain new knowledge for the sake of applying it to practical ends, he will be free only to the extent to which he may choose his field of application without dictation from others.

This absolute meaning of freedom in science is obviously subject to some severe limitations in practice. Such freedom is available to the solitary self-supported scientist, but to scarcely anyone else. Henry Cavendish, the wealthy eighteenth-century bachelor who lived a life of seclusion and had the means to indulge his scientific curiosity in many different directions both in chemistry and physics, provides an interesting example. But he carried his freedom so far that much of his work was not known or appreciated during his lifetime and was rediscovered by others later. Such highly individualized and private science is of course the exception and hardly to be often realized in practice, though other examples might be cited, even in recent times. However, the typical scientific investigator in the university is almost never in this happy situation of being able to indulge his fancy without limit. Modern science is too expensive. Moreover, the professor is expected to devote some attention to teaching, and the time he may devote to research is restricted. This has a definite effect on the research he will choose to work on. However, within this limitation he still possesses enough freedom to choose his own research projects to justify us in referring to him as "free," as far as his science is concerned. J. B. Conant prefers the term "uncommitted" to describe the state of the university professor who picks a research subject of his own free will because his curiosity is aroused and his imagination sees future possibilities in it and not because some superior asks his attention to it as part of a "program." The really severe limitation on the "uncommitted" university scientist is, of course, the financial one. University budgets do not, as a rule, provide liberally for basic research, especially in the costly branches of science. Consequently, support usually must be sought elsewhere and nowadays this means government. So the investigator who wants to pursue an elaborate program must approach an agency like the National Science Foundation and hope that his project proposal will find favor among the referees. If it is too bizarre or unconventional, it probably will not, especially if the would-be investigator is a young and comparatively unknown man. In such a case, freedom becomes illusory, until the researcher somehow manages to establish himself by a lucky breakthrough of sufficient dimensions to attract attention. It may be observed that there is no essential difference between the situation of the would-be uncommitted scientist here and that of the venturer in any free-enterprise project. People who want to get their ideas accepted in any field of human activity must expect to take some risks and see much effort go down the drain. It is not clear why scientific activity should be shielded from this danger.

So it would seem that the alleged precious character of freedom in science may be exaggerated and it doubtless has been by some people. Yet, as we think about it further, we cannot help feeling that the very plentifulness of government financial support of scientific research in the universities may serve almost unconsciously to channel effort into the "fashionable" lines of research rather than encourage the bold and imaginative innovator. The only protection against this danger is the presence on the boards and panels of the government agencies of remarkably enlightened and broad-minded scientists with lots of money at their disposal. Perhaps the United States is wealthy enough and has enough unprejudiced scientists to permit this. There appears to be some fear that it might not work out so well in a country like Great Britain.[4]

On net balance it must be confessed that the worries expressed by some scientists[5] that government support would jeopardize the freedom of scientists in the universities have not been realized. It is true that the usual amount of governmental red tape is involved in the financial accounting of research funds in the universities. But the scientists usually manage to shrug this off. After all, to them all financial accounting is apt to be tedious anyway and better left to business officers. As long as classified contracts are avoided, no problem of freedom of publication arises, and many defense agencies award nonclassified contracts permitting basic research. Moreover,

---

[4] See Bernard Lovell, N. F. Mott, and Hans Krebs, "University Science in Danger," *New Scientist, 12,* 407 (November 16, 1961).

[5] Notably in the United States by the late P. W. Bridgman, who wished to set up an American branch of a British society for freedom in science in the mid 1940's.

the National Science Foundation has a very extensive fellowship program to assist young college and university science faculty members to improve their teaching and research status by study at other institutions both in the United States and abroad. Here again, no strings are attached to the subjects investigated as long as panels of reputable scientists approve of them. Some such appraisal is obviously needed to avoid the financial support of cranks. This to be sure raises the embarrassing question: who is a crank? From an ideal standpoint, this is an unanswerable question. The man with a fixed idea just possibly *may* turn out to be right. In fact, Julius Robert Mayer, who was considered by many a crank on the subject of heat, nevertheless is now admitted to have made a valid calculation of the mechanical equivalent of heat and is considered as one of the founders of thermodynamics. As a practical matter, however, most cranks can be detected, and it is certainly no violation of freedom of science to refrain from encouraging them with liberal financial support.

The problem of freedom of science in industrial research laboratories is a somewhat different one from that in the universities. Industries exist not merely to serve the public through the goods they produce but also to make money for their stockholders. In every industrial laboratory it therefore becomes important that all scientific discoveries be examined to see whether they may lead to patentable devices. It is common to hold up publication of scientific results until this examination has been completed, and then arrange that the publication should not reveal too much to a possible competitor until a patent has been secured. The matter is not a simple one, and it would be quite wrong to give the impression that the right of free publication is denied to the industrial scientist. Yet the freedom is not quite so all-embracing as that of the university scientists. Some concerns are obviously more enlightened than others and effectively give some of their basic scientists almost complete freedom to publish as they wish, but it is in general in lines of endeavor not too closely connected with presumptive patentable inventions. Most large industrial laboratories are anxious that their members should attend scientific meetings, give papers and exchange views with other scientists. They even encourage them to write books (witness the celebrated Bell Telephone Laboratories series) but often demand that the royalties shall accrue to the company instead of the author, a practice not in general employed by universities.

We have already mentioned the security arrangements in defense research, and these obviously enter as a factor governing the scientific activities of investigators in government laboratories, where so much work is closely connected with military requirements. Here, open publication is out of the question, unless declassification takes place, usually a slow and cumbersome process. Yet in spite of this handicap much valuable basic research is carried out in and published openly by government scientific laboratories in the United States, even in those maintained by defense agencies. One can hardly resist the conclusion that though security regulations impose obstacles in the way of scientific progress in governmental laboratories, these have more of a nuisance character than fundamental significance in destroying the essential freedom of their scientific workers. In other words, the situation does not reflect a conscious effort to restrict the freedom of science in accordance with some ideology which considers that science to secure the support of the state must remain subservient to the latter and restrict its attention to projects believed to benefit the state in some practical fashion. It seems unlikely that in the United States such a view will ever take hold as long as the general concept of individual freedom embodied in the Constitution prevails.

On the other hand, the status of science in countries subject to totalitarian regimes would appear to be decidedly equivocal. The history of scientific research in the Soviet Union, for example, in the years since the Russian Revolution of 1917, is not clear, but what is known suggests that freedom in science has suffered many fluctuations. The attempt in the early days of the Communist regime to make scientists toe the party line produced hypocritical acquiescence among some and outright rebellion among others, some of whom unfortunately disappeared in the struggle. The keen realization on the part of the governing body of the dependence of the welfare of the state on applied science gradually gave protection to certain groups because they were too valuable to persecute even if they believed in such heresies as the theory of relativity, which the authorities could not reconcile with dialectical materialism, the official philosophy of the Communist party. Even here, the (to Western

SCIENCE AND THE STATE

scientists) strange episode of Lysenko's rise to fame and power on the resurrected biological theory of the inheritance of acquired characteristics reflects a subservience of science to a particular philosophical doctrine upheld dogmatically by the state. With the death and discrediting of Stalin, freedom of discussion in this field has gradually been restored, but one gets the impression that biological scientists can never have under the Soviet system any real assurance of complete independence of thought.

The physical scientists and mathematicians would appear to be in somewhat happier circumstances. The practical applications of their research are so important for the state that it is better for the government to let them alone as long as they show no desire to meddle with politics. Moreover, it is much easier for these scientists to pay lip service to the official creed of the state, while completely ignoring it in practice. This state of affairs will naturally become more settled as the state becomes more certain of its domestic stability. Hence, it seems probable that the physical scientists, engineers and mathematicians have effectively on the average as much freedom to pursue their investigations as in the Western nations. It is true that they do not have quite the same range of opportunity to attend international meetings outside of Russia, but their visits have increased considerably in the past ten years. The importation of foreign scientific literature into the Soviet Union is very extensive and Russian scientists have no difficulty in keeping abreast of work done in other parts of the world. The high quality of their own work is admitted and admired everywhere. If Western governments had ever indulged the hope (for political reasons) that the presumptive restriction on the freedom of science in Russia would hamper the development of both pure and applied science there, it appears now that this hope was vain.

In our discussion thus far we have taken for granted the assumption that the scientist looks upon his freedom to investigate as a cherished right conferred on him by society. We ought not to leave the subject without having a look at the question: how does the man in the street feel about this freedom? After all, in his capacity as a taxpayer, he is being called on to an increasing extent to foot the bill for the support of this scientific adventuring, and it is only natural that he should wonder at times what this precious freedom of the scientist really amounts to and what it may lead to in the life of society. If he is a socialist, it is clear he will feel that the freedom of the scientist really should mean the obligation to serve the state so as to produce a better life for all citizens. He is not therefore impressed by the plea that the best scientist is the uncommitted scientist; rather, he wishes to see the efforts of the scientist directed toward the practical needs of people. During and after World War II there grew up in Great Britain a considerable body of opinion favoring this point of view, with scientists like the physical chemist J. D. Bernal and science writers like J. G. Crowther[6] associated with it. These individuals took the stand that science developed in the first place largely to meet social need and that its further evolution should favor the social welfare and not merely satisfy the curiosity of the chosen few. This attitude naturally aroused opposition and led to considerable public activity on the part of certain scientists,[7] who have taken the stand that in the long run science will make a greater contribution to human welfare if the state makes no attempt to direct research along so-called practical lines. They evidently feel rather strongly, indeed, that the very strength of the scientific spirit will be sapped if not lost entirely if too much emphasis is laid on the scientist's duty to direct his attention to immediate human needs. The more one ponders on this problem, the more inclined one is to the belief that there is no solution which will satisfy everyone. The whole subject is shot through and through with value judgments to which people cling with vehemence, as indeed they are fully entitled to.

Even the person without socialistic leanings and with no particular love for the welfare state may well have misgivings about the freedom of the scientist to let his investigative urges roam where they will. The average citizen is beginning to understand the significance of the vanishing frontier between basic science and technology, and although he welcomes all the increases in his comfort which technological advances have produced, he worries a good deal

[6] See, for example, J. G. Crowther, *The Social Relations of Science* (London: Macmillan, 1941).

[7] See, for example, John R. Baker, *Science and the Planned State* (London: George Allen and Unwin, Ltd., 1945). Dr. Baker is a biologist. Among his associates in the attack on the socialistic conception of science are people like the physical chemist M. Polanyi, and the biologist A. G. Tansley.

about the compensatory penalties he may be called upon to pay in the dangers inherent in the fearful weapons created by this same technology. Though he may be sensible enough to lay the responsibility for the possible misuse of these weapons on the politicians who play with the destinies of nations, he probably cannot forget that it is the scientists who have made it possible for the politicians to be more dangerous than ever, simply because they have powers conferred by modern science that were not available to their predecessors. Who can foresee the consequences of a growing feeling of irritation over the dangers of science not merely on freedom in science but on the future of science itself?

The forebodings expressed here might be dismissed as mere childish vaporings were it not for the fact that prominent scientists are being asked more and more frequently to give counsel to government on the very highest level. No matter what one may feel about the intrinsic devotion to truth of the scientist, moving in high political circles must impose a severe strain on the disinterestedness of the scientist. Of course, one may take the view that once a scientist has reached this stage he is really an administrator or administrative consultant and the problem of his freedom as a scientific investigator has become illusory. Yet his influence on the freedom of others may be considerable and he may well come to represent science in the eyes of the general public and form the picture the latter will be inclined to act upon.

Thus we see that the problem of freedom of science is a very complicated one. It is a part of the subject matter of what may be called the sociology of science.

CHAPTER 9

SCIENCE AND

HUMAN BEHAVIOR

Human Behavior and Ethics

It will not be denied that the crucial problem in civilization as we know it is the behavior of man toward his fellow man, particularly with respect to those aspects usually referred to as moral conduct. We should not shirk an examination of the relation of science to this problem, which arises since man does not live by himself as an isolated biological entity, but must exist somehow alongside other human beings. As a matter of fact, we have already had to face this situation in our earlier considerations, first and foremost in what is probably the chief anxiety now existent about the progress of science, namely, that it is giving man enormously increased power over his physical environment without at the same time conferring on him the wisdom to use this power for the well-being instead of the harm of his fellow human beings. This is closely connected with the cultural lag stressed on page 88. It is a rather involved and complicated affair, since the whole advancement of science itself is of course a social phenomenon and can hardly be considered without reference to the interactions of people on various levels of sophistication. An aspect of science recognized as of increasing importance is its sociology, dealing with such questions as the cooperative activity of scientists in large-scale research programs, the publication of research results, methods of retrieval of scientific information, the administration of funds for research projects and research fellowships, the communication of the significance of science to the general public, etc. In all these activities, scientist meets with scientist at a level on which understanding and straightforward dealing are essential; at the same time, in many of them the scientist confronts the layman in the endeavor to "sell" the importance of science to the latter. In all these social relations it is commonly assumed that the scientist by virtue of his profession adopts a high moral tone and deals with his associates in accordance with the highest ethical principles: after all, he is a searcher after "truth" and an adherent to "fact." The actuality of this picture is open to some doubt! The only reason we have conjured it up is to emphasize that scientists like other human beings must have relations with their fellows and that to prevent them from being chaotic must follow some rules of conduct usually hallowed by custom and followed on the average in stable societies, even though continually violated by isolated individuals. Such systems of rules when reflected on and refined by philosophers become codes of ethics.

The question we now have to raise is this: what does science have to do with ethics? The question may be handled on two levels. In the first place, we may consider the scientist as a would-be objective observer of the actual behavior of people and as a user of the scientific method to describe and explain this behavior. Sociologists and social psychologists carry out this program, and although, as we have observed in Chapter 5, there are immense difficulties in attaining the same degree of thoroughness and precision as in the natural sciences, yet it is possible to describe human behavior while preserving some reasonable approximation to the scientific method. We shall discuss this in somewhat greater detail in the next section, particularly with reference to conduct deemed right or wrong and about which the thoughtful individual raises the everlasting question: "What ought I to do?" This is one aspect of the bearing of science on ethics: the scientist looks around and observes how in fact people do behave with respect to what they consider to be the most important issues of their individual lives. He may indeed do somewhat more than this. He may try to construct a theory to "explain" moral conduct in the same sense in which a natural scientist explains the experience called external nature through theories from which may be derived by logical deduction laws describing the regularities in this experience.

But this is only one level of the problem. The rather more fascinating one to which some scientists have recently been giving atention is this: can ethics as a set of principles governing moral conduct be developed as a science? If so, what kind of science is it and how is it related to the natural science we have been discussing in this book? Can ethics in this sense be treated as description, creation and understanding of experience? If the purpose of ethics is to provide a guide to conduct, it would appear to be something quite different from science. But one can certainly concoct ethical theories which have the same logical structure as scientific theories. Thus, one can begin with a set of notions about human conduct like honor, duty, justice, love, mercy, and the like, and from these build more precisely specified concepts relating to individuals faced with particular circumstances. Certain hypothetical ethical precepts can then be set up, as, for example, the Golden Rule, Kant's categorical imperative, or the more specific commandments in the Decalogue. From these principles we can seek to derive regularities in human moral conduct under various circumstances which are consistent with the hypotheses. These will be statements that under such and such circumstances, if the principles prevail, a human being will behave in such and such a fashion. Let us take a simple example. Suppose an isolated community of individuals who have precisely the same mental and physical attributes, and further suppose that this means that they all are endowed with the same instincts of self-preservation and self-development. If the Golden Rule is now adopted as a fundamental principle and added to the above special assumption, it follows logically that no person in the community under consideration will ever lay violent hands on another.

Now this is of course a highly and somewhat absurdly idealized train of reasoning, having but a remote connection with actual human experience; but it seems nevertheless appropriate to call it an example of ethical theory. Its relation to science lies essentially in the analogy between its logical structure and that of the standard scientific theory which we have examined in this book. It would be idle to pretend, however, that the analogy is complete. In the scientific theory the aim is to invent a picture out of which follow by logical deduction laws which can be identified with regularities observed in ordinary experience, and a theory has no degree of significance or success unless this identification ensues. In the ethical theory, on the other hand, the purpose is to conclude from certain hypothecated principles of conduct how the individuals in a given population *ought* to behave. To concoct such theories can be an interesting game, but it is a moot question to what extent those seeking a closer connection between science and ethics can be satisfied by it. On the other hand, those who feel strongly that statements which contain the word "ought" have no place in science, i.e., that science has nothing to do with *normative* principles, will by the same token conclude that such considerations as we have just been indulging in are meaningless and a foolish waste of time. Is this then the end of the story and can we salvage nothing?

We shall see later in this chapter that we can indeed provide interesting and possibly fruitful scientific analogies of ethical principles, particularly in the domain of thermodynamics. But before we embark on this, it will be desirable to look a bit further into the scientific description of human behavior as it actually exists. We ought, moreover, in this context to examine some of the ethical problems that are involved in the progress of science and its associated technology.

The Scientific Study of Human Behavior

There seems little point in belaboring the truism that human behavior is enormously complicated. We have had occasion in Chapter 5 to note the difficulties associated with the attempt to describe human social phenomena in scientific terms. Yet people have never ceased to try to account in some fashion or other for the way other people react to the various features of their environment, both living and nonliving. For ages this has been considered the proper province for literary endeavor, whether expressed in secular or religious terms. Human conduct suggests at once the involvement of the humanities, particularly in the face of the apparently inexplicable problem of good and evil in human life. Yet, in more recent times, scientists in the shape of psychologists have not hesitated to step in and examine the human individual with minute care to see what can be plausibly asserted about what makes him tick. In their task they have been joined by a legion of sociologists, anthropologists, theologians and philosophers, all trying to make some sense out of the strange things that human beings do, particularly in what is commonly called moral conduct.

One essential difficulty is clear at the outset. Natural science, as we have discussed it in this book, has achieved success largely because the scientist has been willing to focus attention on relatively small pieces of experience: he has abstracted from the totality of experience decidedly limited domains for special study. In this analytical approach the hope is that by understanding what goes on in the little, and piecing the various domains together, one can gradually enlarge the scope of scientific description. In the case of the human organism, however, it is not clear that this method will be so successful, at any rate where so-called mental reactions are concerned. The argument here runs somewhat as follows: One can study with success the behavior of parts of the human nervous system more or less as a physicist studies the behavior of an electronic circuit. However, the attempt to integrate the knowledge of the various parts into a prediction of the behavior of the whole organism controlled by the nervous system has so far failed. It seems as if one is forced to study the organism as a whole in order really to understand its behavior. But this is a task of enormous complexity if approached by the usual scientific techniques. This is doubtless the basis for the feeling often expressed that the scientific description of the conduct of man is an impracticable program. In spite of this, there has grown up a large body of psychological literature actually describing human behavior. In fact, one school of psychologists considers that the chief task of psychology is just to observe and describe the behavior of man and animals. Its most extreme form, as presented by J. B. Watson, has been given the name *behaviorism*.[1] Though classical introspective psychology felt the need of postulating perception and feelings in animals in the form usually termed instincts, the behaviorist dismisses such assumptions as unnecessary and insists that the study of the animal organism be limited to observed stimuli and the responses which the organism makes to them. The theoretical explanation of the observation is given in terms of reflexes to which the organism becomes conditioned by a learning process. We recall in this connection the famous experiment

[1] J. B. Watson, *Behaviorism* (London: Kegan Paul, 1925).

of I. P. Pavlov on the conditioning of a dog to salivate on the ringing of a bell. The thoroughgoing behaviorist has no use for the concept of instinct as a built-in and inherited tendency of the organism
to behave in certain ways from the very beginning of its life, and
feels that all its behavior is really learned. This extreme view of
Watson and his followers has indeed been vigorously attacked by the
supporters of the instinct theory,[2] who point out that learning and
the building up of reflexes take time, whereas many animals display
characteristic behavior patterns from birth. It is not our purpose to
enter into an evaluation of these rival psychological points of view.
Our main concern is to note that psychologists do consider it worthwhile to describe the behavior of living organisms, including man,
whether in a purely empirical fashion or by the invocation of certain theoretical concepts like instincts. There *is* a science of human
behavior.

It may be objected that we are getting off the track here. For most
of the behavior which psychologists, sociologists and anthropologists
study is not examined from the standpoint of its moral or ethical
implications, and in this chapter we are supposed to be concerned
primarily with the relation between science and ethical behavior. It
is perfectly correct that psychologists devote a lot of attention to
the way the human individual perceives things, i.e., to the nature of
his experience; they study vision, audition, taste and smell and have
added enormously to our understanding of these processes. Ethical
considerations do not enter into such studies any more than they
do into the physicist's explorations into the transmission of light and
sound. If human individuals could be completely isolated, such a
standpoint might be relevant, though even here ethical questions
may arise, since to be sure even hermits have to make decisions how
to treat their own bodies. But certainly as soon as the relations between two or more human beings come into question, we cannot
ignore the ethical aspects of their behavior, no matter what the
ostensible character of this behavior may be. It has been observed by
many[3] that associated with every external stimulus to the nervous

[2] See, for example, Ronald Fletcher, *Instinct in Man* (London: George Allen
and Unwin, Ltd., 1957).
[3] For a fairly recent analytical study see, for example, Nicolas Rashevsky,
Mathematical Biology of Social Behavior (University of Chicago Press, 1957),
especially Chapters 5 and 14.

system of a human individual there is a reaction either pleasant or unpleasant, a feeling of either satisfaction or dissatisfaction. Some mathematical biologists believe indeed that this reaction can be measured, since a given person is prepared to say that one stimulus gives greater satisfaction than another, and hence a crude quantitative scale can be set up. Rashevsky develops this viewpoint in his book and in particular examines in detail the case in which an individual acts deliberately in such a way as to maximize his total satisfaction from all stimuli. Rashevsky naturally refers to this as "hedonistic" behavior, reflecting the philosophical doctrine that pleasure is the highest good. Observation indicates that it is sufficiently widespread to deserve careful attention. For our purposes it is illuminating to employ it to interpret the interaction of two individuals A and B. Let us suppose that A has experienced a certain amount of satisfaction from some external stimulus and expresses it by some overt behavior, e.g., laughing or weeping (note that satisfaction can be *negative*). This behavior acts as a stimulus on B, producing in him a certain amount of satisfaction. Now A may not care anything at all about B and may be concerned only with his own satisfaction. In maximizing the latter he may be said to act *egoistically* (Rashevsky's terminology). On the other hand, if B's satisfaction is positive under the stimulus of A's behavior, when A strives to maximize his own, he presumably increases that of B. In this case, Rashevsky refers to A's act as *altruistic*. Presumably, his behavior is altruistic whether or not he wishes consciously to increase B's satisfaction. We well know there are people who go around shedding sweetness and light for the satisfaction it gives them without giving a thought to other people! Still another possibility is that the behavior of A in maximizing his own satisfaction may produce negative satisfaction in B and indeed increase its absolute magnitude. In this case, Rashevsky refers to A's behavior as *sadistic*. Presumably, this terminology will apply whether A's action is deliberate or not: thus one may have unconscious sadism, though it is not customary to accept this in real life.

Rashevsky develops some ingenious mathematical analysis to describe both egoistic and altruistic behavior in the case of the interaction of two or more individuals. Some of this is based on the work of the psychologist and authority on psychometrics L. L. Thurstone.[4] It is not our intention to pursue this in detail. It will suffice to emphasize that psychologists do concern themselves with human behavior involving value judgments and indeed moral values, and moreover consider that such judgments can be assessed quantitatively. It is inevitable that such studies as those we have mentioned will be subjected to criticism, but there also seems no reason to doubt that they will continue to be made and will achieve a measure of success.

The studies of men like Rashevsky and Thurstone are comparatively recent. We must not conclude, however, that the attempt at a scientific description of human ethical behavior is a new development. As a matter of fact, in Western civilization it goes back to Plato and Aristotle, and all subsequent philosophers have had a crack at it in one way or another. But we do not wish to review ethics as a branch of philosophy. Let us take a look at what the biologists of the 19th century have done with it. They were of course swayed by the evolutionary ideas which came to a kind of climax with the publication of Darwin's *Origin of Species* in 1859 and his *Descent of Man* in 1871. In the latter work, the great naturalist extended his theory of evolution by natural selection to man and in particular stressed the role played by the development of ethical ideas as a part of organic evolution. He lays great emphasis on the value of altruism for the success of the tribe, i.e., the sacrifice of the individual for the sake of the group to which he belongs. These views were examined and extended by such people as Herbert Spencer, who in his *Principles of Ethics* (1892) elaborated the view that good conduct on the part of the individual is that which leads to a satisfactory life in society, whereas bad conduct is that which makes for unhealthy and disruptive social relations. Spencer's ideas undoubtedly influenced many, though not all the evolutionists followed him. T. H. Huxley, for example, though a vehement propagandist for Darwinism, finally decided he could not accept evolutionary ethics. In his famous Romanes lecture of 1893 on "Evolution and Ethics" he states his mature conclusion that the altruistic motive is in direct contradiction to the aggressive motive indispensable for success in the struggle for existence, and that there can never be a

[4] See, for example, Louis L. Thurstone, *The Measurement of Values* (University of Chicago Press, 1959), in particular Chapter 12.

genuine equilibrium between these two motives. Though Darwin appeared to accept the Golden Rule as the basis for morality on an evolutionary ethics, Huxley could not see it. He felt its weakness is that all people simply do not have the same nature or attitude toward human behavior: why should one pamper the criminal whose only response will be to turn and rend you? The other cheek when turned usually gets slapped! Without doubt, Huxley would have considered his judgment confirmed by the present population explosion associated with what many conscientious people look upon as the proper ethical treatment of so-called underprivileged communities. Here, certainly, is part of the ethical dilemma of science which we shall discuss in some detail in the next section.

Perhaps Huxley overlooked some bright spots in his generally grim appraisal. Thus, for example, in the treatment of human beings by other human beings the race *has* succeeded in eradicating actual slavery in so-called civilized communities. This must be considered ethical progress on any theory which assigns dignity to the human individual; it represents an evolution of what may be termed better *average* behavior. Admittedly, there are statistical fluctuations. Cruelty of individual to individual and group to group unfortunately continues to exist under the best of conditions and of course has reached almost unbelievable depths of savagery in wartime. An objective biologist like Sir Charles Sherrington[5] has felt it necessary to stress that the "will to live," which apparently characterizes all living things, implies in turn the essentially unavoidable cruelty of life. All our experience teaches us that life has to exist at the expense of other life. For nourishment we are all condemned to live off our relatives, in so far as we must think of all living things as constituting one great unity. Even vegetarians cannot evade this, since vegetables share life with animals. Hence, cruelty is somehow built into our life experience. It is not merely the deliberate cruelty of the organism with the highly developed nervous system. As Sherrington points out, the lowly malaria parasite goes through its life cycle with apparently no appreciation of the cruel devastation it is wreaking in human beings. It too is, so to speak, determined to live even at the expense of more elaborate organisms. When we decide to stamp it

[5] C. S. Sherrington, *Man on His Nature*, 2d ed. (Cambridge University Press, 1951).

out with all the means at our disposal are we displaying a proper "reverence for life" or are we cruel also?

When we meditate on such things, we are probably inclined to think that after all Huxley took the right tack and that we have little ground for believing that a satisfactory system of ethics has evolved in the development of man from lower forms of life. As reflective human beings, in our optimistic moments we may be moved to feel that life is the crowning blessing of our planet, but we can hardly blame the pessimist for seeing with what appears to him to be equal justice this self-same life as the greatest curse afflicting the earth. Sherrington sums up his own attitude toward altruism by voicing the opinion that it really involves sharing the suffering of others as if it were our own. For many, this is a hard road to follow.

The Ethical Dilemma of Science

Here we return for the moment to questions like those raised in the first chapter of this book. Is there not a real danger that though the role of science in society may be highly beneficial as far as the material wants of man are concerned, the ultimate influence on the moral behavior of people may be deleterious? The problem is a many-sided one. In our first look at it at the beginning of the book, we emphasized what many consider the more crucial aspect, namely, the very real danger that the vast powers put in man's hands by science-inspired technology may shortly lead to the destruction of vast numbers of the population of the earth and the breakdown of the civilization to the building up of which technology has so notably contributed. This is indeed a dilemma and an anxious one, for self-preservation is one of the most powerful of human instincts. The result is fear, and it may well be influencing the mental health of many persons whose adjustment to life is rather delicately balanced.

We expect to go into this further, but at the moment should like to consider a more subtle facet of the dilemma, not so much concerned with fear of sudden death brought about by the abuse of scientifically conferred powers, but rather with the fear of the inevitable and remorseless erosion of human values by a change in the equilibrium of human existence. Let us take a look at this.

The development of moral principles in Western civilization has been a long and involved process, but no one can doubt that much of this has been closely associated with the primary necessity of human labor to sustain existence. This was indeed erected into a religious principle by the authors of the Old Testament when they made God announce to the sinners in the Garden: "In the sweat of thy face shalt thou eat bread." Those who have lots of work to do have little time for vice. It has been observed by W. E. H. Lecky[6] that slavery in the later Roman Empire exercised a corrupting influence on the free population, since it deprived them of the necessity for labor. They became lazy, interested only in the free food distributed by the state and the entertainments provided by the gladiatorial contests. And along with this went a breakdown in common morals of daily life. Since human slavery has disappeared in Western civilization, the relevance of this to our theme of the possible role of science and technology in the erosion of modern ethical values may seem slight. What we must remember, however, is that though man in the twentieth century has no human slaves to work for him, he has countless energy slaves. We have commented on this earlier. To an accelerating extent the physical labor of individuals is being replaced by energy supplied by the various sources mentioned in Chapter 7 and made available for human purposes by a myriad of energy transformers—the numberless gadgets that now surround the lives of every one of us. Even the labor of thinking is being taken over by the enormously rapid large-scale computing machines. The inevitable result has been a very great increase in human leisure in so-called civilized communities. Along with this has come a decided decrease in the effort required to meet nearly all the normal necessities of life. Not only has transportation been revolutionized by the motor car, but people no ,onger have to exert an effort to transfer themselves over the smallest distances: walking as a means of transportation in the United States is rapidly disappearing. All domestic arrangements connected with cooking, heating, cleaning, etc., have been taken over by the electrical "energy slaves." At the same time, cheap "entertainment" has multiplied enormously, and it is entertainment which makes few demands on the mental

[6] W. E. H. Lecky, *History of European Morals from Augustus to Charlemagne*, vol. 1 (New York: Appleton and Co., 1895), Chapter 2.

activity or any physical effort (more than that involved in flicking a switch) on the part of the spectator. Printed matter abounds, but with its increasing use of pictures makes fewer and fewer demands on the mental effort of the reader. All along the line the amount of energy expenditure on the part of the human individual in Western civilization required for daily existence appears to be decreasing. The inevitable question arises: is this a good thing? There is no reason to believe that we shall ever answer it! But this in turn is no reason why we should not consider it. Many thoughtful people undoubtedly feel that modern technology is "softening" the population both physically and morally through its emphasis on materialistic values and through its denigration of labor, both physical and intellectual. That this is an extreme and highly prejudiced view is maintained by others who point out with some justice that the very technology inveighed against has demanded for its development much intellectual labor. Moreover, along with increased leisure has come emphasis on the pursuit of hobbies which demand increasing attention and provide an outlet for superfluous energy. The only sensible course is probably to reserve judgment on the whole matter, though there are times when one cannot help wondering what the breakdown in civic morale will be in the face of a dire emergency like nuclear warfare, when one observes how easily normal activities are brought to a standstill in an average-size city by a severe snowstorm or even the warning of an approaching hurricane.

Let us return to the ethical problems involved in the possible misuse of atomic energy. The fact that the first large-scale application of atomic fission was the killing of vast numbers of people has naturally perturbed many, who feel that it would have been better had the possibility of the release of energy on this scale never been discovered. So the dilemma at once appears: if scientists were to refrain from seeking to create new experience in the so-called atomic domain, they would deprive human beings not only of the satisfaction of a deeper understanding of nature but also of the possibility of replacing the gradually disappearing energy sources from fossil fuels with new and practically inexhaustible supplies. On the other hand, the pursuit of such a goal when successful at once exposes mankind to the danger of extinction by nuclear explosion or serious bodily harm and probable genetic damage through radiation fallout resulting from the explosions. One can take the stand that the scientist faces a very difficult ethical decision. However, closer inspection seems to show that this sort of ethical dilemma is really involved in practically every activity indulged in by man: no matter what he does or achieves, someone can abuse it to the damage of others. All food is poison when consumed to excess. It is therefore unreasonable to single out the scientist as providing a unique ethical dilemma.

A. V. Hill, the British physiologist, thinks that if it is meaningful at all to talk about the ethical dilemma of science, the most serious example is the population explosion. The dilemma here is easily expressed. The vast increase in scientific knowledge in both the biological and physical sciences, especially in the past seventy-five years, has put in the hands of medical people methods for the control of disease of almost fantastic efficacy. The normal ethical ideas of both Western and Eastern civilizations insist that these methods shall be applied; in the name of common humanity we can do no less. The result has meant the saving of countless lives, especially among the young, which had hitherto fallen a sacrifice to disease. But the result has been a striking increase in the population and hence more mouths to feed. In Eastern countries the struggle to find sufficient food to maintain populations increasing at accelerated rate is a difficult if not in many instances a hopeless one. One is therefore entitled to ask: is it really ethical after all to save all those lives only to allow their possessors to be afflicted with malnutrition or even starvation?

The proponent of the unrestricted use of science and technology in modern society is not after all helpless in the face of this particular dilemma. He can think of two possible scientific attacks on the problem. Assuming first that natural fertility is allowed to prevail with corresponding birth rates, the task confronting the technologist is the increase in the food supply and other facilities for civilized living. In the heavily industrialized West this challenge has been met and the production of adequate food and housing is not a major problem. In Eastern countries not so well favored by climatic conditions and less advanced industrially the technological task is truly formidable. This is not to say that it is hopeless. Certainly India is making progress, though the situation in China seems distinctly less favorable. The real catch comes in the first premise that the birth rate can be stabilized. Thomas Robert Malthus did not think so when he wrote his famous *Essay on the Principle of Population* in 1798. He was sure that the population can always be depended on to crowd up to the food supply, for the latter can increase only arithmetically, while the former increases geometrically unless artificially checked. The Malthusian doctrine lost favor in the face of the great economic and industrial development of the nineteenth and early twentieth centuries, but contemporary thought is again veering in its direction. Sir Charles G. Darwin in his book *The Next Million Years* (referred to in Chapter 5) can see no way to ignore it safely. There remains, however, another possible scientific solution to the dilemma: artificial control of the birth rate. There has been a great deal of experimental study of contraceptive devices during the past fifty years with the result that reasonably reliable chemical birth control devices are now available, though their cost on a large scale is not insignificant. There seems to be no doubt that science can solve the problem of birth control completely and ultimately at a price society can afford. At the moment, the greatest obstacle appears to be the religious one. Obviously another ethical problem is involved. Deeply religious people may easily convince themselves that tampering with the natural process of the reproduction of the race is highly unethical. If we accept this point of view, the fundamental dilemma is still with us.

It may be objected that the population problem is not quite so simple as we have just presented it. It is possible that a given nation may wish its population to increase at an accelerated rate for the sake of the presumptive gain in military power. To feed these teeming millions it will then overrun its neighbors and ultimately envisage world domination. Leaving aside ethical considerations and even scientific ones for that matter, this plan based purely on power politics is apt to be a risky business, though it is just conceivable that an already overpopulated country under a totalitarian regime might in desperation try it.

To return to the question of the ethical dilemma of science, all indications are that it will not be resolved in any of its phases by abandonment of scientific and technological attacks on human problems. A method which has been found so successful will not be lightly abandoned. If there is anything after all to the idea that ethical principles have evolved along with the growing up of the human race, it is conceivable that the progress of science will gradually modify these principles in the direction of providing for the future health and safety of mankind. This suggests another look at the scientific status of ethics. Can it be developed as a science?

Can There Be a Science of Ethics?

Having made clear that human behavior, and in particular that usually called moral, can be studied scientifically, and having pointed out the dilemma involved in the influence of science and its applications on human conduct, it now remains finally to return to the considerations earlier in this chapter and examine the question: can there be a science of ethics? What shall we mean by this question? We obviously do not mean a scientific study of human conduct as it is actually observed to exist. This we have already studied briefly. Presumably we must mean the formulation of ethics in the form of a scientific theory with its constructs, postulates, deductions and verifications, as set forth in Chapter 2. It is clear that the deductions from any ethical theory must be in the nature of normative statements that something *ought* to be so and not merely descriptive statements that something *is* so. The problem then is: what kind of constructs and postulates will lead to such deductions?

This question has recently been studied in some detail by Henry Margenau,[7] and in this section we shall follow the broad outline of his treatment with special comment where appropriate. The first requirement is to establish the character of the postulates of ethics. Margenau refers to these as "axioms," but since we have used the term postulate so extensively in our discussion of the nature of scientific theory, it is proper that we should continue our use of it. It is true that the term axiom in common parlance conveys the notion of self-evident truth, while postulate usually connotes a more speculative assumption. In this sense, the choice of the word postulate seems more appropriate in order to bring out the essentially hypothetical character of the assertions in question. Now the postu-

[7] Henry Margenau, personal communication (1962) of material ultimately to appear in a book.

lates of a scientific theory are statements about the nature of the fundamental constructs of the theory or relations connecting two or more of them. We recall that one of the postulates of the molecular theory of gases is the existence of molecules and their motion in a closed space, as well as that their average behavior entails the large-scale observed properties of the gas in question. One of the fundamental postulates of the theory of mechanics is the assumed relation connecting the acceleration of a particle with its mass and the resultant force acting on it. The fundamental postulate of the electromagnetic theory of light is that light is simply the propagation of an electromagnetic field through space and time. This involves the assumptions about the relations connecting the electric and magnetic field intensities which are known as the Maxwell field equations. All these postulates assert something in sentences which are indicative in character. They ask us to base our further conclusions on hypothetical statements that certain things *are*.

On the other hand, if there is to be such a thing as an ethical theory, its fundamental postulates will necessarily be statements that the basic entity in the whole theory, namely, the human being, *ought* to do so and so. They must therefore be *imperatives* or commands, like the Golden Rule, the Ten Commandments or Kant's categorical imperative ("Act only on that maxim whereby you can at the same time will that it should become a universal law"). No one can question the enormously significant role which such commands have played in the development of the life of man on the earth. Certain people have been forever ordering others to do thus and so, and the others have made various responses depending on their reverence or lack of it for the person uttering the command. Governments have made obedience to command the basis for the stability of society. The word "command" has ever had a primary meaning in military activities. Small wonder that such imperatives should have served as the basis for ethics in practically all religions.

There is one point that might provoke puzzlement in this connection. Is there something so uniquely characteristic of the imperative that its content cannot be equally well expressed in terms of the indicative? Certainly such a statement as "Thou shalt not kill" can be translated into the equivalent indicatives "It is wrong to kill," or "You ought not to kill anyone," or "A person who kills is subject to punishment." The same kind of translation is available for every imperative. In each case, a value judgment makes its appearance with the consequent weakening of the emphasis that is inherent in the command. Linguistically, the indicative form may be equivalent to the imperative, but it certainly lacks the force of the latter. We can of course take the stand that the value judgment expressed explicitly in the indicative form is implicit in the corresponding imperative, and that a person who responds not emotionally but logically to such statements may be as much inclined to question the imperative as the indicative. This is apparently part of the price we must pay for the privilege of casting our ethical system in the scientific mold. The postulates must be transformed from indicative value judgments to commands, which have a stronger emotional emphasis, or if we wish to say so, a greater moral force. It may not be too clear scientifically what this means, but let us in any case proceed on this basis. We must indeed be careful not to conclude that because commands involve emotional and moral appeal they therefore lose contact with logic. It is expected that one can draw conclusions from them just as in natural science one makes deductions from the postulates of a theory.

It is in order to raise the query: where do the imperatives or commands of an ethical theory come from? We discussed in Chapter 2 the corresponding problem for scientific theory and found that scientific hypotheses have various origins. Some are definitely and rather directly suggested by experience while others are products of a vivid imagination working on experience. We had to admit that fundamentally they all possess an arbitrary character, which indeed serves to color scientific theories in a way not always accurately recognized by the casual observer. The arbitrariness, indeed, is not always appreciated even by the inventor of the theory, since with all the intensive thought that goes into the creation of a successful theory there is apt to go along an assurance that the hypotheses finally adopted are somehow self-evident. Their essential arbitrariness will, however, forcibly strike the independent examiner who has not been thinking along the lines of the inventor. Gradually, as the theory gets accepted and enshrined in the body of scientific thought, familiarity makes the arbitrary character grow dim and finally almost disappear as a matter of practical concern.

Thus, the success of mechanics as a physical theory has all but eliminated any suggestion of arbitrariness about Newton's famous "laws" of motion from the minds of the mechanical engineers.

How do matters stand, then, with the commands of ethical theory? The commonly accepted view is that they have resulted from the collective wisdom of the human race, based on experience. This has much to commend it, and yet the fact that the commands are so often disobeyed suggests that after all they do have an arbitrary aspect. Many have emerged from the minds of thoughtful people who evidently felt that human relations would be better were such commands obeyed. Ethical theory like scientific theory seems to have originated in the minds of profound thinkers, who fell back for justification not merely on experience but on their own imaginative powers. In the case of ethics, the latter has often been termed illumination or revelation. It can hardly be considered essentially different from the stroke of inspiration at the basis of the invention of a new scientific theory.

Margenau discusses the interesting problem of the relation between ethical imperatives and legal codes and considers that though there are obvious differences, one may plausibly think of the two as connected in much the same way as science and engineering. The ethical command is a general injunction to the individual to behave in a certain way. It presumably applies to all individuals in a given society. A statutory law is a specific injunction to do something or not do something which is supposed to relate to the best interests of society at large. Laws are enacted deliberately by groups of people who have taken it upon themselves or have been delegated to govern others. Ethical commands usually arise from the insight of individuals. Laws provide for punishment for their infraction. Ethical commands provide for no enforcement except through the conscience of the individual, unless indeed the commands form a part of his religion and he is a devout believer. Finally, laws may cover cases in which most persons would not feel that ethical considerations enter very obviously. However, in spite of these differences we may fairly look upon jurisprudence as the effort of society to formalize the enforceable parts of ethical codes and apply them to the behavior of society as a whole—ethical engineering, in short.

The importance of a postulate or set of postulates in a scientific theory resides in what it implies, i.e., in what can be deduced from it bearing some identification with observed phenomena or more particularly with those observed regularities in experience which we represent by laws. We therefore expect that the importance of the imperatives in an ethical theory is inherent in their implications and the relevance of these implications for human conduct. As a matter of fact, there exists a name for the process of drawing conclusions from ethical commands: it is casuistry. The trouble with it is, however, that in the course of time it has come to have a pejorative or depreciative connotation as implying sophistical or specious reasoning, and this indeed is its common significance today. To use it would therefore be to court misunderstanding. We shall therefore refer to the process in question as moral implication or, to follow Margenau, moral explication.

In the first place, the logical consequences of ethical imperatives are human actions which possess certain properties or qualities. These qualities are called values. Let us give an illustration with the Golden Rule as the ethical imperative. But first we must make clear that the use of the word "logical" is not quite like its use in the standard deductions from the postulates of natural science. The "logical" consequence of the imperative is here interpreted to mean the *motivation* of the human action with respect to the attainment of certain values. Let us suppose the value is happiness. Presumably an individual seeks his own happiness as one of the precious values of life. The Golden Rule will then motivate him to seek the happiness of others as the surest way of assuring his own. If indeed his happiness results by virtue of his seeking, then we may say that the ethical theory so constructed has been validated with respect to this value. One can proceed to test the theory with respect to a whole hierarchy of values like love, justice, beauty, etc., not all of which may be, to be sure, appropriate to the particular imperative used as the basis of the ethical system.

The general idea of the method seems clear. It is also evident that it is open to attack on several grounds. An obvious one is connected with the difficulty of precisely defining a particular value so that it will have the same significance to all members of the group supposed to be subject to a given ethical theory. What is happiness, for example? Any statement about this value is of course a value judgment and will differ in general from person to person. The individ-

ual whose chief happiness is fishing and who therefore, to insure
the happiness of others, persists in making them gifts of fishing
tackle may be disagreeably surprised at the results. Of course, the
truly altruistic person will endeavor to ascertain what makes his
fellow man happy and then act accordingly. But this usually presents very great difficulties.

Another difficulty with the theoretical ethical system as presented
is that the process of explication of a moral imperative is by no
means so clear-cut as the purely logical deductions from the indicative postulates of a theory in natural science. This is doubtless due
to the relatively enormous complexity of human behavior and the
difficulty of interpreting it in terms of a definite set of moral values.
Naturally, this is not an insuperable obstacle to the acceptance of a
scientific theory of ethics, since it must also be faced by the social
psychologist in his scientific study of human behavior.

One feels indeed a certain discomfort in trying to attack such a
viewpoint as has been just expounded, for inevitably a scientist is
tempted to look at it from the standpoint of what natural science
tries to do. He is obviously impressed with the success of science as a
predictive scheme and hence as effectively a creator of experience.
This would seem to be outside the ken of an ethical theory. On the
other hand, the ability of a relatively small number of moral imperatives to motivate all the validated behavior of men with respect
to a set of accepted values suggests a persuasive all-embracing possibility which is certainly not to be too lightly depreciated and
reminds one of the descriptive aim of science to subsume under a
small number of constructs and associated postulates a relatively
large body of experience.

Whatever attitude one takes toward the detailed attempt to erect
ethics into a science like natural science, one can at any rate see the
importance of seeking for the most general type of moral imperative, so that a relatively small number of these will be sufficient to
provide satisfactory motivation for successful social living. Here we
find scope for scientific effort on a somewhat different level, namely,
the search for ideas in the natural sciences leading to imperatives
which, because of their construction with general concepts that have
proved very successful in coping with experience from a scientific
point of view, may provide a more suggestive basis for ethical action.
This is a somewhat elusive and probably thankless task. Neverthe-

less, we shall close our labors in this book by studying one such possibility which will form the topic for our final section.

A Scientific Analogy: The Thermodynamic Imperative

In Chapter 6 we paid some attention to the principles of thermodynamics as a very general type of scientific theory, and introduced the concepts of energy and entropy. At that stage our aim was to show the relevance of these concepts to information theory. We are now going to utilize them for a somewhat different purpose. It will be recalled that whereas the first principle of thermodynamics expresses the constancy of the energy of an isolated physical system and by extension that of the energy of the whole universe (Clausius), the more subtle and in some ways more frustrating second principle expresses the inevitable tendency of the entropy of the universe to increase. This means that in spite of the constancy of the energy, the opportunity to transform this energy into useful work steadily diminishes. The steady production of entropy is the natural and prevailing process throughout the universe. In order to provide a plausible basis for understanding this rather mysterious second principle or law, which in the first instance emerged from the rather obstinate refusal of heat engines to function at 100 per cent efficiency under any attainable circumstances, we had recourse to a statistical interpretation. From this point of view, the second law means merely that physical systems naturally tend to change from less probable to more probable states. Finally, we showed that this is equivalent to saying that the natural tendency for systems is to move from states of order to states of disorder, and that any attempt to reverse this process involves the uncompensated expenditure of energy. According to the second law, the final state of the universe (if it has any such!) is one of complete randomness or disorder.

We noted that there *are* examples of local entropy consumption. Some of these are contrived by man, as in devices like mechanical refrigerators, for which we pay by entropy production elsewhere. The principal natural entropy consumption in human experience is exhibited by the living organism, which in a fashion only imperfectly understood as yet transforms disorder into order by building out of a few chemical elements compounds of great complexity but well-defined order; these form the constituents of the living cell, out of which in turn the various parts of the organism are created. In other words, life is a natural consumer of entropy on a local scale. In the face of the general increase in the entropy of the universe as a whole this, as was pointed out in Chapter 6, may well be considered of major importance in any overall interpretation of living things and their relations to each other as well as to their whole environment. As we saw, local consumption of entropy is not to be considered a genuine violation of the second law, for it seems altogether likely that the entropy consumption of living things is compensated for by the corresponding entropy production elsewhere in the universe. Nevertheless, it contains a powerful suggestion that man as a creature with a highly developed nervous system has an obligation to act deliberately as an entropy consumer to put himself in tune with the very principle of life which he and his existence represent. As a matter of fact, man's whole struggle to introduce order into a chaotic environment may be seen as a kind of intuitive recognition of this obligation. He builds dwellings rather than live in the open; he develops means of transportation; he cultivates the soil rather than rely on what nature provides without his efforts; he develops language to put regularity into communication with his fellow men. Practically every element in man's developing civilization may be interpreted either as an instinctive or conscious and deliberate attempt to replace disorder with order, in other words to consume entropy.

It will, of course, be objected that we have vastly overstated the case in our eagerness to stress an interesting thermodynamic analogy. The picture is not indeed as simple as we have painted it. There are obvious fluctuations in the consumption of entropy by living things. We recognize that destructive tendencies are exhibited by many human beings, and to this extent they are entropy producers rather than consumers. We think at once of arsonists and murderers, the former destroying by fire what was carefully and slowly constructed by human labor and thus producing a vast increase in entropy in a short time, and the latter reducing in an instant the highly complicated but very orderly arrangement of molecular constituents we call a human individual to the equivalent of a heap of disorderly rubble with a similar enormous relative increase in entropy. We call such people criminals and enemies of society, precisely because they sin against the aim of man to produce and maintain order in his affairs. On a less tragic scale, it is not without significance that we attach the word disorderly to the alcoholic and the drunkard. They are nuisances to society in that they produce more than their fair share of the disorder in a social milieu which strives for entropy consumption. Everyone can think of other illustrations of such nuisances. Even the thoughtless litterbug who strews the landscape with his rubbish is in his lowly way an unnecessary entropy producer.

But in spite of these melancholy examples, man in his better moments seems to exemplify a ceaseless urge to force some order on his experience. The very existence of science is an example of this. This reflects a conscious desire and involves thoughtful planning limited, to be sure, to a certain small fraction of the population but implying consequences for all mankind of the sort we have been examining in this book.

Considerations of this kind seem to suggest a new kind of imperative which if reasonably interpreted might serve as a satisfactory basis for an ethical code. We shall call it the *thermodynamic imperative* and phrase it as follows: All men should fight always as vigorously as possible to increase the degree of order in their environment, i.e., consume as much entropy as possible, in order to combat the natural tendency for entropy to increase and for order in the universe to be transformed into disorder, in accordance with the second law of thermodynamics.

Let us now undertake an examination of this imperative, its possibilities and drawbacks.[8] It will be simplest to do this in terms of the various objections that either have been raised or might be raised against its value and significance. One obvious one is that even in the case of living things, the inexorable increase in entropy

[8] For the introduction of the name and a preliminary presentation, cf. R. B. Lindsay, "Entropy Consumption and Values in Physical Science," *American Scientist, 47,* 376 (1959). For further discussion, development and criticism, cf. William Malamud, "Psychiatric Research: Setting and Motivation," *The American Journal of Psychiatry, 117,* 1 (1960). Also, S. Polgar, "Evolution and the Thermodynamic Imperative" *Human Biology, 33,* 99 (1961); H. S. Seifert, "Can We Increase Our Entropy?" *American Scientist, 49,* 124A (1961). See also the correspondence by J. S. Kirkaldy, Anthony Standen and others in *American Scientist, 47,* 326A, 398A (1959). For an earlier popular account of the relation between entropy, order and life see E. Schrödinger, *What is Life?* (New York: Macmillan, 1945), Chapter 6. See also E. M. Fournier D'Albe, "El Problema Etics del Cientifico," *Cuadernos del Seminario de Problemas Cientificos y Filasoficos,* No. 10. Second Series (National University of Mexico, 1955).

goes on, the second law always wins in the end. All experience points to the fact that every living organism eventually dies. Death is associated with the reduction of the highly developed order of the living organism to a random and disorderly collection of molecules. We are reminded that we are "dust" and to "dust" we ultimately return. So some may take the stand, what is the use of fighting this inevitable trend? If there is something in the universe which prefers the steady transition from order to disorder, why should we try to combat it? Why should we not adjust ourselves to the irresistible order of events, get on the toboggan and enjoy the ride? This view, while doubtless satisfactory to the out-and-out hedonist, seems to overlook the significance of man's long, hard struggle to preserve the gift of life and exploit its possibilities. To maintain life in the face of the terrible challenge of the environment must have been an ordeal to our primitive ancestors. They responded to the challenge by ignoring the processes of decay and by endeavoring to establish order out of the chaos of their surroundings. If they had failed to meet this challenge, we would obviously not have been here today to discuss the question of our own obligation. It is a plausible historical theory, vigorously promoted by Arnold Toynbee,[9] that civilizations owe their success to the extent to which they meet the challenge of the environment. Now certainly the second law of thermodynamics provides the greatest overall challenge to man's existence and progress on earth. The suggestion is strong that no matter who wins in the end we ought to oppose the second law to the best of our ability: this is the essential intent of the thermodynamic imperative.

Before we go on to further objections, and indeed to enable us to appreciate the force of another objection to the imperative, let us consider an interesting aspect of the principle, namely, its possible quantitative character. To understand this, it is necessary to recall some of the thermodynamic developments in Chapter 6. There we emphasized the representation of the entropy of a system in a given state as a constant times the logarithm of the statistical probability, which in turn is measured by the number of ways in which the given state can be realized. The larger the number of ways, the more random or disorderly the state, and conversely. Hence, in the last analysis we should be able to measure relative order merely by

[9] Arnold Toynbee, *A Study of History*, Abridgment of vols. 1-6 by D. C. Somervell (Oxford University Press, 1945).

counting, the simplest of all mathematical processes. To be sure, the practical difficulties are immense in the case of systems of some degree of complexity, like living organisms and collections of organisms. But the important thing is that in principle it is possible, and this to the quantitatively minded should prove an attractive feature of the thermodynamic imperative. Even the practical difficulties need no longer appear hopeless in the face of the availability of more and more elaborate large-scale computers, which can carry out arithmetical processes with enormous speed.

There are of course the qualitatively minded folk (the vast majority, it must be confessed) who deem it repugnant and even absurd that one should base our ethics on an imperative whose implications can be worked out by such a thoroughly impersonal process as counting. They will be inclined to raise the question: just what is this order which we are asked to strive to increase and how does it show itself? It is one thing to say that a crystal exemplifies a high degree of order, whereas the same atoms or molecules if allowed to roam around freely, as in a gas, constitute a state of much less order or what may properly be called a disorderly or random state. But it would appear to be quite another matter to say that we can measure order in living organisms, human beings and human relations in the last analysis merely by counting. This objection has great force and to meet it demands a more careful investigation of the concept of order than we have given hitherto or can hope to give in this introductory presentation. The subject is undoubtedly complicated and to a certain extent puzzling. We can, however, offer some tentative and preliminary thoughts on it, ignoring the quantitative aspect.

It has been objected that a literal adherence to the thermodynamic imperative can lead to absurdity. For, since living things consume entropy locally, the best way to insure maximum entropy consumption and production of order in our surroundings is to promote the production of living things, presumably vegetable life as well as the lower animals and man. Now, since these various forms of life live off each other, it is clear that entropy consumption of one form can take place only at the expense of entropy production in the sacrifice of the others. Man naturally sees himself in the role of the primary entropy consumer. The conclusion one might then be tempted to draw from the imperative is to encourage the production of as much human life as possible. But this would lead to various types of problems (including what in conventional society would be regarded as immoral behavior, e.g., free love, etc.), the worst of which would undoubtedly be that overpopulation which we have seen is connected with one of the most serious ethical dilemmas of science.

There exists a way to meet the objections stressed in the preceding paragraph. There are more ways than one of skinning a cat, and the process of entropy consumption and order building is not confined to the reproduction of life, though that may be considered Nature's way of assuring that the second law does not have things all its own way on a local basis. But too much emphasis on this unconscious entropy consumption should not blind us to the conscious building of order in man's environment by the deliberate exercise of human thought and effort. As we have already pointed out, man in his long evolution has been forever striving to build order into his experience both with respect to the external environment and to his relations with his fellow human beings and other living things with which he has to share his earthly home. This is well brought out by S. Polgar in his article on "Evolution and the Thermodynamic Imperative," already mentioned earlier in this section. He points out that there is some plausibility in the view that the thermodynamic imperative itself may have undergone an evolution "from the unconscious adaptations of infrahuman organisms, through the development of 'exosomatic' culture [i.e., culture not directly connected with bodily reactions as such but associated with mental activity—a concept stressed by A. J. Lotka in 1945: see the bibliography in Polgar's article] by which the great stores of information can be transmitted to subsequent generations and otherwise distributed, to the scientific descriptions of evolution and the deliberate attempts to increase the availability of energy for survival and the enjoyment of life." All around us we find evidences of the way in which man has built order into his life. Many of the resulting institutions we tend to take for granted, e.g., the banking houses, insurance companies, educational institutions, but they are all a part of the ceaseless search for order in social relations. What the thermodynamic imperative asks is that every individual human being do his part through his own life to add and not to subtract from this order—to consume as much entropy as he can. For the average individual this will scarcely demand more than the following of the ethical precepts of the Golden Rule and Kant's categorical imperative in all aspects of his personal and social life. For the unusually talented person, however, it does impose correspondingly heavy responsibility, for by very virtue of his superior ability he has it in his power to be either an unusually great producer of entropy or an unusually great consumer. Clever people can contribute mightily to the increase in the order in man's environment, as the labors of scientists and engineers have amply demonstrated. On the other hand, when they take the form of unscrupulous political manipulators, their activities can lead to chaos. The situation here is a bit subtle. A political leader with power over the lives of millions of people may well delude himself with the notion that in introducing order into their lives, and by gaining control over millions more, he is effectively an entropy consumer on a grand scale. It is conceivable that a dictator like Hitler thought of himself in this light. But directed entropy consumption of this sort hardly satisfies the needs of the thermodynamic imperative, which cannot be effective for the individual unless he is free. All imperatives of ethical value must rest on the assumption that the person on whom the injunction is laid has the power to obey or not as he wills. There can be no meaningful ethical code for a population of robots or automatons. The idea of the thermodynamic imperative has indeed been attacked on the basis of its ultimately leading to a social strait jacket. Certainly, if the maximization of social order means a system where the behavior of every human being is rigidly specified and controlled like the positions of the atoms in a crystal (which as we have seen does exhibit a very high degree of order and hence a minimization of entropy) most intelligent people would find it utterly distasteful. But we repeat the thermodynamic imperative applies to the individual and has no meaning unless he is free to obey it or disobey it, as he sees fit.

The imperative has some interesting implications about life itself. We have seen that the existence of living organisms is an example of local, so to speak, unconscious entropy consumption. We may plausibly look upon the evolutionary process which has produced Homo sapiens as a living creature with a highly developed nervous system as an effort to attain the highest degree of order compatible with the peculiar properties of the stuff out of which living things are made. Nevertheless, we are persuaded that the evolutionary process still has a long way to go and that the possibilities for still greater order or entropy consumption to be built into man are very great indeed. However, our experience seems to show also that the pace of the process is painfully slow. In the replication of life, we are bound at one stage of the game by fairly restrictive hereditary laws, and are more or less at the mercy of chance mutations to produce any striking change in fundamental human characteristics. It is therefore all the more encouraging that Nature has after all provided us with opportunities to consume entropy in a conscious, deliberate fashion through the exercise of our wits and indeed to devise new types or order which the instinctive method of entropy consumption would never have realized.

In this connection we may find some worthwhile ideas in the writings of the British biologist P. B. Medawar,[10] who has stressed that in planning for the continuation of life, Nature has not always made favorable genetic arrangements, particularly with respect to immunological devices. Though life itself is a great entropy consumer, there are some aspects of it connected with genetics which are by no means propitious for the maximization of order. This places on the human mind all the greater responsibility to seek out in experience those aspects conducive to order. Suggestions along this line are continually being made by scientists and technologists, though not always with entropy consumption ostensibly in mind. A recent excursion into this field is that of J. S. Seifert, whose *American Scientist* article has been referred to earlier in this section. He finds much that is disorderly in our educational system, for example, and particularly in the ways in which we try to train scientists and engineers, and visualizes a fertile field for the creation of more order in this segment of society. It is obvious, of course, that each thoughtful person will have his own ideas how to make our educational system more orderly. This is another illustration of the fact that the precise definition of order in society is not easy. But are we thereby relieved of the obligation to do whatever we can to relate our human fate to the thermodynamic measure of order through the concept of entropy?

[10] P. B. Medawar, *The Future of Man* (London: Methuen and Co., Ltd., 1960). See also his *The Uniqueness of the Individual* (London: Methuen and Co., Ltd., 1957).

We regret we are unable through any mystic insight to give the average man any more elaborate recipe for following the thermodynamic imperative. If indeed he does his best to follow the Golden Rule and the categorical imperative of Kant, as well as to display as much reverence for life as is consistent with staying alive, he will probably not go far wrong in offering the optimum challenge to the second law. But there is perhaps something else he can do. As one who feels that it is only through the exercise of his intellect that man will ultimately cope successfully with the problems that confront him, he should do all in his power to see that intellect is cultivated by all those who have it in more than average measure. This means he should encourage in every possible way the arts and science. In particular, he should make it his business to understand the role of science in society, to support its beneficent applications and to resist its abuses. Since it is the fundamental purpose of science to find order in our experience and of technology to apply and maintain it in all the goings-on of our daily lives, to take the thermodynamic imperative seriously means to foster by all possible means the progress of science. In stressing this we do not intend to neglect or depreciate the arts. Fournier d'Albe, in the article on "Values and the Scientist," referred to earlier, mentions with approval the assertion that "the function of art is to introduce order into emotional experience." With this we thoroughly agree. This book will have failed of its purpose if it has not made clear the close relations connecting science and art in all its phases, as well as the humanities in general. The "house of intellect" should be one house. From this point of view the challenge of life is the better understanding of life and its purpose. For some this implies the search for forms in which to clothe our emotional reactions. For others it means the search for concepts with quantitative, numerical measure to provide an orderly mold into which to place as much of our experience as possible. Since we are continually creating new experience, the forms and measurable concepts will change correspondingly as the ages pass. But we cannot help believing that certain concepts like the thermodynamic ones of energy and entropy are so general that we can make them serve for a long time to come. This is why the thermodynamic imperative has seemed so attractive as a contribution of science to the problems of human life.

BIBLIOGRAPHY

Chapter 2. What Is Science?

1. W. S. BECK, *Modern Science and the Nature of Life* (London: Macmillan, 1957).
2. P. W. BRIDGMAN, *The Logic of Modern Physics* (New York: Macmillan, 1927).
3. P. DUHEM, *The Aim and Structure of Physical Theory* (Princeton University Press, 1954).
4. J. HADAMARD, *Psychology of Invention in the Mathematical Field* (Princeton University Press, 1945).
5. R. B. LINDSAY AND HENRY MARGENAU, *Foundations of Physics* (New York: Dover Publications, Inc., 1957).
6. HENRY MARGENAU, *The Nature of Physical Reality* (New York: McGraw-Hill Book Company, 1950).
7. A. MOLES, *La Création Scientifique* (Geneva: Éditions René Kister, 1957).
8. H. POINCARÉ, *Foundations of Science* (Tr. by George Bruce Halsted. Lancaster, Pennsylvania: The Science Press, 1946).
9. C. SHERRINGTON, *Man on His Nature* (Cambridge University Press, 1957).
10. E. SCHRÖDINGER, *What Is Life?* (New York: Macmillan, 1945).
11. R. TATON, *Reason and Chance in Scientific Discovery* (London: Hutchinson, 1957).
12. A. N. WHITEHEAD, *Adventures of Ideas* (Cambridge University Press, 1939).

Chapter 3. Science and the Humanities

1. G. D. BIRKHOFF, *Aesthetic Measure* (Cambridge: Harvard University Press, 1933).
2. SIR R. BRAIN, *The Nature of Experience* (London: Oxford University Press, 1959).
3. D. BUSH, *Science and English Poetry* (London: Oxford University Press, 1950).

4. R. G. COLLINGWOOD, *The Principles of Art* (London: Oxford University Press, 1938).

5. B. I. EVANS, *Literature and Science* (London: George Allen and Unwin, Ltd., 1954).

6. E. FAURE, *The Spirit of the Forms* (New York: Garden City Publishing Co., 1937).

7. R. P. GRAVES, *Life of Sir William Rowan Hamilton*, 3 vols. (Dublin University Press, 1882-1887).

8. H. v. HELMHOLTZ, *Popular Lectures on Scientific Subjects,* Second Series (New York: Appleton, 1881).

9. J. W. KRUTCH, *Human Nature and the Human Condition* (New York: Random House, 1959).

10. E. MacCURDY, *The Notebooks of Leonardo da Vinci* (New York: Garden City Publishing Co., 1941).

11. J. REDFIELD, *Music—A Science and an Art* (New York: Tudor Publishing Company, 1926).

12. C. P. SNOW, *The Two Cultures and the Scientific Revolution* (Cambridge University Press, 1959).

13. P. WIENER AND A. NOLAND, *Roots of Scientific Thought* (New York: Basic Books, 1957).

Chapter 4. Science and Philosophy

1. R. CARNAP, *The Logical Syntax of Language* (London: K. Paul, Trench, and Trubner, 1937).

2. E. CASSIRER, *Substance and Function* (LaSalle, Illinois: The Open Court Publishing Co., 1923).

3. C. J. DUCASSE, *Philosophy as a Science* (New York: Oskar Piest, 1941).

4. A. S. EDDINGTON, *The Philosophy of Physical Science* (New York: Macmillan, 1939).

5. A. S. EDDINGTON, *Relativity Theory of Protons and Electrons* (Cambridge University Press, 1936).

6. P. FRANK, *Between Physics and Philosophy* (Cambridge: Harvard University Press, 1941).

7. P. FRANK, *Philosophy of Science* (Englewood Cliffs, New Jersey: Prentice-Hall, 1957).

8. N. R. HANSON, *Patterns of Discovery* (Cambridge University Press, 1958).

9. A. O. LOVEJOY, *The Great Chain of Being* (Cambridge: Harvard University Press, 1936).

10. C. S. PEIRCE, *Principles of Philosophy, Vol. 1 of Collected Papers* (Cambridge: Harvard University Press, 1931).

11. J. ROYCE, *The Spirit of Modern Philosophy* (Boston: Houghton Mifflin Co., 1892).

12. B. RUSSELL, *A History of Western Philosophy* (New York: Simon & Schuster, Inc., 1945).

13. R. SCHLEGEL, *Time and the Physical World* (Michigan State University Press, 1961).

14. S. TOULMIN, *The Philosophy of Science* (London: Hutchinson University Library, 1953).

15. G. J. WHITROW, *The Natural Philosophy of Time* (London: Nelson, 1961).

Chapter 5. Science and History

1. H. ADAMS, *The Education of Henry Adams* (Boston: Houghton Mifflin Co., 1918).

2. W. BAGEHOT, *Physics and Politics* (New York: Knopf, 1948).

3. G. BARRACLOUGH, *History in a Changing World* (Oxford: Basil Blackwell, 1955).

4. J. D. BERNAL, *Science in History* (London: Watts, 1954).

5. L. DE BROGLIE, *The Revolution in Physics* (New York: The Noonday Press, 1953).

6. H. T. BUCKLE, *History of Civilization in England*, "The World's Classics," 3 vols. (London: Oxford University Press, 1903).

7. H. BUTTERFIELD, *The Origins of Modern Science* (London: G. Bell and Sons, Ltd., 1950).

8. E. H. CARR, *The New Society* (London: Macmillan, 1953).

9. W. C. D. DAMPIER-WHETHAM, *History of Science* (New York: Macmillan, 1930).

10. C. G. DARWIN, *The Next Million Years* (London: Rupert Hart-Davis, 1952).

11. H. DINGLE, *The Scientific Adventure* (London: Sir Isaac Pitman and Sons, 1952).

12. H. A. L. FISHER, *History of Europe*, 3 vols. (Boston: Houghton Mifflin Co., 1935).

13. E. GIBBON, *The History of the Decline and Fall of the Roman Empire*, with notes by H. H. Milman, 2 vols. (n.d.) (London: Ward, Lock and Co.).

14. A. R. HALL, *The Scientific Revolution, 1500-1800* (London: Longmans, Green and Co., 1954).

15. J. MADGE, *The Tools of Social Science* (London: Longmans, Green and Co., 1953).
16. J. T. MERZ, *A History of European Thought in the Nineteenth Century*, 2 vols. (Edinburgh: Blackwood, 1904).
17. O. NEUGEBAUER, *The Exact Sciences in Antiquity* (New York: Harper Torchbooks, Harper & Brothers, 1962).
18. G. PEACOCK, *Life of Thomas Young* (London: John Murray, 1855).
19. K. R. POPPER, *The Poverty of Historicism* (London: Routledge and Kegan Paul, 1957).
20. C. SINGER, *A Short History of Science* (Oxford: Clarendon Press, 1941).
21. W. H. WALSH, *Philosophy of History* (New York: Harper Torchbooks, Harper & Brothers, 1960).
22. A. WOOD, *Thomas Young, Natural Philosopher* (Cambridge University Press, 1954).
23. G. K. ZIPF, *Human Behavior and the Principle of Least Effort* (Cambridge: Addison-Wesley, 1949).

Chapter 6. Science and Communication

1. W. R. ASHBY, *An Introduction to Cybernetics* (London: Chapman and Hall, Ltd., 1956).
2. E. T. BELL, *Men of Mathematics* (New York: Simon & Schuster, Inc., 1937).
3. L. BRILLOUIN, *Science and Information Theory* (New York: Academic Press, 1956).
4. C. CHERRY, *On Human Communication* (New York: Technology Press of M.I.T. and John Wiley and Sons, 1957).
5. I. B. COHEN AND F. A. WATSON (editors), *General Education in Science* (Cambridge: Harvard University Press, 1952).
6. J. B. CONANT, *On Understanding Science* (New Haven: Yale University Press, 1947).
7. J. B. CONANT, *Science and Common Sense* (New Haven: Yale University Press, 1951).
8. T. DANTZIG, *Number, The Language of Science* (New York: Macmillan, 1954).
9. C. DARWIN, *The Expression of the Emotions in Man and Animals* (New York: Appleton, 1873).
10. H. FLETCHER, *Speech and Hearing in Communication* (New York: D. Van Nostrand, 1953).
11. G. T. GUILBAUD, *What Is Cybernetics?* (London: Heinemann, 1959).

12. G. H. HARDY, *A Mathematician's Apology* (Cambridge University Press, 1941).
13. T. H. HUXLEY, *Science and Education—Essays* (New York: Appleton, 1902).
14. P. DE LATIL, *Thinking by Machine—A Study of Cybernetics* (London: Sidgwick and Jackson, 1956).
15. C. W. MORRIS, *Signs, Language and Behavior* (New York: Prentice-Hall, 1946).
16. D. PEDOE, *The Gentle Art of Mathematics* (London: English Universities Press Ltd., 1958).
17. W. V. O. QUINE, *Methods of Logic* (London: Routledge and Kegan Paul, 1952).
18. E. G. RICHARDSON, *Technical Aspects of Sound,* Vol. 1 (Amsterdam: Elsevier Publishing Co., 1953).
19. B. RUSSELL, *Principles of Mathematics* (New York: W. W. Norton and Co., Inc., 1903).
20. M. SCHLAUCH, *The Gift of Language* (New York: Dover Publications, Inc., 1955).
21. A. TARSKI, *Introduction to Logic and the Methodology of the Deductive Sciences* (London: Oxford University Press, 1946).
22. A. R. UBBELOHDE, *Man and Energy* (London: Hutchinson, 1954).
23. S. ULLMANN, *Principles of Semantics* (New York: Philosophical Library, 1951).
24. N. WIENER, *Cybernetics* (New York: John Wiley and Sons, 1948).
25. B. L. WHORF, *Collected Papers on Metalinguistics* (Dept. of State, Washington, D. C., 1952).

Chapter 7. Science and Technology

1. E. ASHBY, *Technology and the Academics* (London: Macmillan, 1958).
2. E. AYERS AND C. A. SCARLOTT, *Energy Sources, the Wealth of the World* (New York: McGraw-Hill Book Co., 1952).
3. J. G. CROWTHER, *The Social Relations of Science* (New York: Macmillan, 1941).
4. T. K. DERRY AND T. I. WILLIAMS, *A Short History of Technology* (Oxford University Press, 1961).
5. L. A. HAWKINS, *Adventure into the Unknown—The First Fifty Years of the General Electric Research Laboratory* (New York: Morrow, 1950).
6. H. S. W. MASSEY, *Atoms and Energy* (New York: Philosophical Library, 1956).

7. C. E. K. Mees, *The Path of Science* (New York: John Wiley and Sons, 1946).

8. P. C. Putnam, *Energy in the Future* (Princeton, N. J.: D. Van Nostrand Co., 1953).

9. J. Read, *Through Alchemy to Chemistry* (London, G. Bell and Sons, Ltd., 1957).

10. H. Thirring, *Power Production* (London: George G. Harrap and Co., Ltd., 1956).

11. G. P. Thomson, *The Forseeable Future* (Cambridge University Press, 1955).

12. A. R. Ubbelohde, *Man and Energy* (London: Hutchinson, 1954).

13. A. Wolf, *History of Science, Technology and Philosophy in the 16th and 17th Centuries* (London: George Allen and Unwin, Ltd., 1935).

14. A. Wolf, *History of Science, Technology and Philosophy in the 18th Century* (New York: Macmillan, 1939).

Chapter 8. Science and the State

1. J. R. Baker, *Science and the Planned State* (London: George Allen and Unwin, Ltd., 1945).

2. J. F. Baxter, *Scientists Against Time* (Boston: Little, Brown & Co., 1946).

3. J. G. Crowther and R. Whiddington, *Science at War* (New York: Philosophical Library, 1948).

4. J. S. Dupré and S. A. Lakoff, *Science and the Nation* (Englewood Cliffs, N. J.: Prentice-Hall, 1962).

5. P. M. Morse and G. E. Kimball, *Methods of Operations Research* (New York: Technology Press of M.I.T. and John Wiley and Sons, Inc., 1951).

6. J. Perry, *The Story of Standards* (New York: Funk & Wagnalls Co., 1955).

7. H. D. Smyth, *Atomic Energy for Military Purposes* (Princeton University Press, 1945).

8. E. N. Titterton, *Facing the Atomic Future* (London: Macmillan, 1956).

Chapter 9. Science and Human Behavior

1. E. W. Barnes, *Scientific Theory and Religion* (New York: Macmillan, 1933).

2. E. A. Burtt, *Religion in an Age of Science* (New York: Henry Holt and Co., 1929).

3. R. FLETCHER, *Instinct in Man* (London: George Allen and Unwin, Ltd., 1957).
4. M. B. HESSE, *Science and the Human Imagination* (London: SCM Press, Ltd., 1954).
5. T. H. HUXLEY, *Evolution and Ethics and Other Essays* (New York: Appleton, 1898).
6. W. JAMES, *Varieties of Religious Experience* (New York: The Modern Library, n.d.).
7. W. E. H. LECKY, *History of European Morals from Augustus to Charlemagne*, Vol. 1 (New York: Appleton, 1895).
8. P. B. MEDAWAR, *The Future of Man* (London: Methuen and Co., Ltd., 1957).
9. P. B. MEDAWAR, *The Uniqueness of the Individual* (London: Methuen and Co., Ltd., 1957).
10. J. E. NEFF, *Cultural Foundations of Industrial Civilization* (Cambridge University Press, 1958).
11. N. RASHEVSKY, *Mathematical Biology of Social Behavior* (University of Chicago Press, 1957).
12. C. S. SHERRINGTON, *Man on His Nature* (Cambridge University Press, 1951).
13. E. W. SINNOTT, *Biology of the Spirit* (London: Victor Gallancz, Ltd., 1956).
14. H. SPENCER, *The Principles of Ethics,* two vols. (New York: Appleton, 1899).
15. L. L. THURSTONE, *The Measurement of Values* (University of Chicago Press, 1959).
16. A. TOYNBEE, *A Study of History,* abridgment of vols. 1-6 by D. C. Somervell (Oxford University Press, 1945).
17. C. F. VON WEIZSÄCKER, *The History of Nature* (University of Chicago Press, 1949).
18. J. B. WATSON, *Behaviorism* (London: Kegan Paul, 1925).
19. A. N. WHITEHEAD, *Science and the Modern World* (Cambridge University Press, 1933).

INDEX

Fossil fuels, 227 f.
Foucault, Jean Bernard Léon (1819–1868), French physicist, 30, 51
Frank, Philipp (1884–), Austrian-American physicist and philosopher, 87
Free will, 93
Freedom of science, 262 ff.
French encyclopedists, 71
Frequency in acoustics, 54
Frequency, rank order relation in language, 171 f.
  of sound wave, 137
Fresnel, Augustin Jean (1788–1827), French physicist, 121
Froude, James Anthony (1818–1894), British historian, 111
Fucks, Wilhelm, contemporary German physicist, 176 f.
Fundamental theorem of algebra, 195

Galilei, Galileo (1564–1642), Italian physicist, 2, 33, 54, 117, 207 f., 239
Gauss, Karl Friedrich (1777–1855), German mathematician, astronomer and physicist, 119, 194
Gene theory of heredity, 22
General Electric Company Research Laboratory, 216
General Motors Laboratory, 219
Geometry, laws of, 26
Germer, Lester Halbert (1896–), American physicist, 221 f.
Gibbon, Edward (1737–1794), British historian, 102, 104, 111 ff., 247
Gibbs, Josiah Willard (1839–1903), American physicist, mathematician and chemist, 29, 31, 213 f., 216
Gilbert, William (1540–1603), English physician and physicist, 118 f., 239
Gödel, Kurt (1906–), Czech–American mathematician, 87, 193
Goethe, Johann Wolfgang, von (1749–1832), German poet and dramatist, 71
Golden Rule, 272, 278, 285, 296, 298
Government, financial support of science by, 255 ff.

Government—(*Contd.*)
  technological and scientific needs of, 235 ff.
Graphic arts, 56 f.
Gravitation, 3
  constant of, 11, 12
  Newtonian hypothesis of, 3
Greek fire, 247
Green, George (1793–1841), English mathematician and physicist, 27 f.
Greenwich Observatory, 240
Gresham's law, 16
Gunpowder, 246

Hadamard, Jacques Solomon (1865–19—), French mathematician, 31
Haeckel, Ernst Heinrich (1834–1919), German biologist, 30
Hahn, Hans (1879–1947), Austrian mathematician, 87
Hamilton, William Rowan (1805–1865), Irish mathematician, astronomer and physicist, 29, 48, 68 ff., 121 f.
Hamiltonian method in mechanics, 82
Hanging Gardens of Babylon, 235
Hardy, Godfrey Harold (1877–1947), English mathematician, 195
Harmonic sound, 137
Harrison, John (1693–1776), English horologist, 240
Heat death, 158
Heat engine, 209 f.
Heaviside calculus, 194
Heaviside, Oliver (1850–1925), English physicist and electrician, 191
Hegel, Georg Wilhelm Friedrich (1770–1831), German philosopher, 101
Helmholtz, Herman Ludwig Ferdinand, von (1821–1894), German physicist and physiologist, 61
Henry, Joseph (1797–1878), American physicist, 215
Henry's law, 16
Hereditary mechanics, 107
Heredity, laws of, 27
Hero of Alexandria (ca. 3d century A.D.), Greek geometer and philosopher, 209
Herschel, Frederick William (1738–1822), German–English astronomer, 96

312